Dynamics of Human Biocultural Diversity

Dynamics of Human Biocultural Diversity

A Unified Approach

Elisa J. Sobo

Taylor & Francis Group

LONDON AND NEW YORK

First published 2013 by Left Coast Press, Inc.

Published 2016 by Routledge
2 Park Square, Milton Park, Abingdon, Oxon OX14 4RN
711 Third Avenue, New York, NY 10017, USA

Routledge is an imprint of the Taylor & Francis Group, an informa business

Library of Congress Cataloging-in-Publication Data

Sobo, Elisa Janine, 1963-
  Dynamics of human biocultural diversity : a unified approach / Elisa J. Sobo.
    p. cm.
  Includes bibliographical references and index.
  ISBN 978-1-61132-190-6 (pbk. : alk. paper)—ISBN 978-1-61132-695-6
    (consumer ebook)
  1. Human ecology. 2. Biodiversity. 3. Cultural pluralism. I. Title.
  GF41.S63 2013
  305.8—dc23
                                    2012033899

Cover design by Lisa Devenish

ISBN 978-1-61132-190-6 paperback

Contents

Preface

In Western society, we tend to think of biology and culture as totally separate and even opposing domains. This book takes a different viewpoint. This book's inquiry into human biocultural diversity takes a connectionist approach. It introduces the general principles of anthropology and specific biocultural topics by focusing on the synergistic interaction of biology and culture, or nature and nurture. In other words, this book demonstrates that we are, simultaneously and self-reinforcingly, biocultural creations.

Intended for entry-level courses in anthropology, medicine, human ecology, and other fields that investigate biology and culture as well as interested readers, this book is unique in its holistic, systems-based approach, and in the way that it uses complexity theory and ideas from the emerging field of epigenetics to shift our thinking about biocultural diversity away from the traditional dualistic or either-or mode. Instead, readers learn to conceptualize human biology and culture concurrently: as an adaptive constellation that exists in relation to the broader environment in which humans live. Our adaptive biocultural capacity has helped to produce the rich range of human diversity seen today.

Human biocultural diversity has long been a key area of study within anthropology. Other fields speak of biocultural diversity, too; applied conservationists, for instance, reference all plant and animal life with the phrase, pointing to the role of human culture in relation to ecological balance and change. Humanity's impact on the earth system has in fact been so great that some scientists have declared we are now in a new geological epoch, the Anthropocene ('anthropos' means people). The human species is itself part of the earth system, and so humans, too, are affected by human cultural action. Indeed, the emergence of human culture complicated our species' evolution, in that the reactive process

7

of natural selection now was conjoined with a proactive helper (culture); this led to greater diversity between human groups than otherwise would have been seen. It is this diversity—human biocultural diversity, and the dynamic, interconnected processes that bring it about—in which I am most interested.

The Biocultural Approach

Until now, many biocultural anthropologists have focused mainly on the 'bio' half of the equation, using 'biocultural' generically, like biology, to refer to genetic, anatomical, physiological, and related features of the human body that vary across cultural groups. The number of scholars with a more sophisticated approach is on the upswing, but they often write only for super-educated expert audiences. Accordingly, although introductory biocultural anthropology texts make some attempt to acknowledge the role of culture, most still treat culture as an external variable—as an add-on to an essentially biological system. Most fail to present a model of biocultural diversity that gives adequate weight to the cultural side of things.

Note that I said most, not all: happily, things are changing. A movement is afoot to take anthropology's claim of holism more seriously by doing more to connect—or reconnect—perspectives from both sides of the fence. Ironically, prior to the industrial revolution and the rise of the modern university, most thinkers took a very comprehensive view of the human condition. It was only afterward that fragmented, factorial, compartmental thinking began to undermine our ability to understand ourselves and our place in—and connection with—the world. Today, the leading edge of science recognizes the links and interdependencies that such thinking keeps falsely hidden. Building on the related campaign for a "new biocultural synthesis" (for example, Goodman and Leatherman 1998), this book takes culture seriously.

This book takes culture so seriously, in fact, that my use of the term 'biocultural' may seem generous in places where I apply it to traits that, on the surface, seem only biological. The same might be said for my use of the term in regard to traits that appear as if only cultural. Yet, our adaptive biocultural capacity operates constantly even when not obviously. This book invites readers to examine that fact. It challenges readers to look beyond shallow definitions of human variation focusing on traits, such as skin color, to include higher-order and more complexly derived variations, such as are seen in the diverse health profiles of various populations. In this book I strive to not simply describe our diverse biologies:

I aim to examine how cultures create and maintain this variation. This includes addressing some of the misfits between modern lifestyles and certain features of the human body as it has evolved to date.

I use no hyphen in the term 'biocultural' because, instead of treating culture and biology separately, the book demonstrates how they really are connected. For example, the book shows that we humans make cultural interpretations of given biological processes and, in turn, manage and thereby modify these processes culturally, including through culturally infused forms of social organization. It describes a variety of ways in which our cultural modes of management affect our physical existence, such as when we settle in a particular geographic region, invent and adopt certain occupational roles, or subscribe to particular social structures, marriage and mating rules, or political systems. All such things alter our bodies in patterned, population-specific ways, much as do more obviously body-related practices such as circumcision or diet. Accordingly, in addition to genetic and related physiological diversity, I include in this book important anthropological subjects where the co-mingling of biology and culture are most obvious, such as kinship and gender. In support of this orientation, I use the term **culture** broadly, as is typical in U.S. anthropology, to refer to the totality of each human group's shared, learned heritage. This includes social, political, economic, religious, cosmological, linguistic, health promotion, and other systems that, taken together, comprise our cultures.

In sum, from this book's perspective, **biocultural diversity** refers to all population-based human variation generated in or reflecting the dynamic, synergistic communion of biology and culture, neither one of which can function without the other. In other words, biocultural diversity, as I use the term, brings together topics that narrowly focused scholars might see as separate; it demands that we acknowledge and investigate their co-constitution. The result of such a unified approach is a book that offers a more comprehensive view on human variation than existing biocultural texts, truly engaging both sides of the biocultural coin and reflecting a holistic anthropological sensibility.

Roadmap

This book is organized into three sections that build incrementally on the ideas of dynamic systems, environmental pressure, natural selection, and adaptation. Chapters in Part I, "A Systems View of Human Adaptation," address our emergence as cultural beings through the process of natural selection and the biological diversity that resulted as humans spread out around the world. Part II, "Socio-political and Economic Factors,"

Figure 0.1 Skull exhibiting cranial modification. *Photo courtesy Valerie A. Andrushko.*

examines the varied adaptations and health profiles that were created around the world as an unintended consequence of the agricultural revolution and related transformations in how groups lived. For example, we examine the spread of infectious disease and resulting changes in the gene pool as well as the way health has become linked to wealth. The last part of the book, "Meaningful Practices," explores our expectations and experiences of our bodies, for instance in the ways in which our different cultures create bodies as male or female or in how standards of beauty lead some groups to skeletal modifications (see Figures 0.1 and 0.2).

Key Features

Health and Diversity

Many of the examples I use in the following chapters relate to health, and there are good reasons for this. First, health snapshots or 'epidemiological profiles' vary from group to group; they are a key indicator and mechanism of diversity. Second, varying health levels serve as an excellent index of adaptation. For instance, ill health suggests that a population is under stress or poorly adapted to its current environment. The focus on health helps students carry forward the initial lesson that biocultural diversity emerged as humans adapted to new or changed environments.

Figure 0.2 Figure 32 from *Le Corset* (The Corset) by Ludovic O'Followell, 1908.

For instance, consider darker-skinned peoples now living in Britain. In comparison to many other places in the world, Britain is an environment with generally low levels of ultra-violet or UV light exposure: the island experiences late sunrises and early sunsets in winter, and often has overcast skies. These factors are compounded by the fact that British society has primarily indoor jobs and many sun-blocking high-rises. These features put darker-skinned peoples living in Britain at a higher risk for health problems such as rickets. In turn, lighter skinned populations relocated to a higher-UV environment (for example, San Diego, California) are at higher risk for skin cancer, sun burn, folate deficiency, and other conditions. Without understanding the interrelatedness of culture and biology—without a comprehensive definition of biocultural such as this book offers—these 'facts' might be described, but their contingent qualities would remain unquestioned and our understanding of the diversity described would remain under-developed at best.

Learner Focused

The integration of "nature and nurture" into a complex systems framework is revitalizing the study of biocultural diversity. Questions asked and lessons learned by the vanguard of scholarship in this area can and should be brought into introductory teaching. To create a rich learning experience that supports such an approach, this book:

- Teaches sophisticated concepts in jargon-free, reader-friendly language that renders them clear and accessible to readers from any discipline;

- Uses student-centered, current, and relevant examples from the past and present that bring lessons to life and demonstrate their here-and-now application;

- Keeps readers on track by focusing each chapter around answering three or so key conceptual questions that could easily be adapted for use as study questions or essay questions on tests.

In addition, each part of the book is introduced with a short essay steering readers toward the main points that the subsequent chapters discuss, and a concluding chapter highlights important connections.

Easily Supplemented

I developed this book over years of teaching biocultural diversity and with input from about two thousand students. Built around lectures fine-tuned over time, it is organized not only for reader ease but also so that it might provide other instructors the backbone for a similar course. Depending on the needs of end users, it could stand alone or be paired with supplemental resources.

In the old days, we might have assigned a reader to complement this text; today, using key terms from each chapter, instructors also can use library databases and the internet (see Appendix and References also) to compile a host of rich, supplementary, up-to-date material that might include both scholarly and popular articles; videos; web-based tutorials; images; podcasts; and so on. We can even engage students actively in creating their own course supplements. For example, using adaptations of Bernie Dodge's inquiry-oriented Webquest method, students can be sent to the internet on a topic- or problem-based treasure hunt. Of course, full-length books also are an option; some that I have used are in References, as are many of the short, topical readings and videos that I have assigned.

Accessible

This book builds on the scholarship of others. However, as this text is introductory, I have not burdened readers with a massive load of references. Whenever possible, I have chosen to cite works that are easy to find and to read; in addition to scholarly literature, I point readers to solid work written for the educated public and published in books, newspapers, educational websites, or public broadcasting sources.

In addition to minimizing in-text citations, the book is only lightly illustrated. This not only limits distractions but also helps keep the book's price relatively low. Accordingly, older photographs and artwork available via the public domain are included, not to suggest that the cultures pictured are dead when indeed they are not, but rather to ensure the book's affordability.

Broad Appeal for Teaching

Instructors with various kinds of expertise in an expanding number of disciplines are charged with teaching biocultural diversity courses. Because of its straightforward approach, this book is likely to appeal to anyone teaching such a course, not only biocultural anthropology specialists. The contents are suitable for other courses as well, from introductory courses in general and medical anthropology to courses that include a review of the rise of civilization and even human biology programs seeking to enhance their culturally informed course offerings. Because this book uses health as a key index of adaptation, it also will fill a niche in public health programs, which increasingly are committed to teaching cultural competence and offering courses that address health inequities and global health.

End Aims

In any course, this book will introduce the amazing—and necessary—diversity of humankind. By the end of the book, readers will be able to define and apply systems thinking, describe the key socio-cultural and political-economic reasons for human biocultural diversity past and present, and characterize some of the many ways in which the human being is, for better and for worse, an irreducible biocultural creation.

Acknowledgments

This book summarizes lectures presented in an anthropology course called "Dynamics of Biocultural Diversity," which I've taught at San Diego State University (SDSU) since joining the department in 2005. Thanks are due to the multitude of students who've taken the course for their enthusiastic interest in the topic and, more so in this context, for their patient, constructive feedback on my methods for explaining and illustrating the fascinating multi-directional links between culture, biology, and human diversity. Without student feedback, this book would not exist. I also wish to thank colleagues at SDSU (particularly Brock Allen and Andrea Saltzman Martin, as well as others too numerous to mention) for providing specific guidance to me in developing the course, first for the lecture hall setting and more recently for online learners. I am extremely grateful for the lessons that they, and my students, have shared with me.

Thanks also are due to Jennifer Collier, for champion editorial oversight. Several anonymous reviewers and Erin Riley kindly provided thoughtful and constructive input on the manuscript and sections of it, and colleagues too numerous to list by name have aided me in my thinking through discussions of topics now reflected in the book's contents. Christopher Morgan graciously fact-checked some of the work, and Jennifer de Garmo helped hunt down illustrations. Dedicated graduate students and teaching assistants Samantha Coppa, Jules Downum, Elizabeth Herlihy, Sonia Khachikians, and Harrison White worked diligently to transcribe lectures, help with figures, copyedit and critique the manuscript, and sharpen the course from which this book grew.

A Systems View of Human Adaptation

Conceptualizing biology and culture separately, as if each exists independently, may be convenient, but it dulls our understanding of who we are, how we work, and how we fit into the world. Part I of this book lays the foundation for a comprehensive and unified understanding of human biocultural diversity—one that embraces the interdependence of biology and culture and all that they entail. Through Part I, readers learn to think holistically, and to see the world from a systems perspective. Part I also introduces students to the concept of adaptation, which is revealed to be a complex process entailing multiple forms of feedback through which the theoretically separate domains of body, behavior, and environment work as one in effecting change.

Chapter by Chapter Overview

Chapter 1 cuts straight to the chase, describing the nature-nurture debate and then demonstrating how it is mistaken and misleading. Rejecting the dualist position, it presents an alternative, unified approach grounded in systems thinking. It explains and exemplifies the holistic viewpoint of anthropology, paying special attention to the concept of adaptation and the phenomenon of emergence. 'Complexity' is defined as a technical term, and explored in regard to how it can inform our understanding of human nature.

Chapter 2 introduces evolutionary biology with a focus on genetic adaptation. After giving readers a good grasp of how genes and traits are passed on to the next generation, this chapter explains how natural selection at the population level works, and why certain forms of genetic variation are useful for a group's trans-generational survival or reproductive success from generation to generation.

In Chapter 3, we learn that the environment plays a key role in the ways that genes are expressed to begin with. This chapter introduces the concept of developmental adjustment, distinguishing it from natural selection and using that as a jumping-off point for an exploration of how genes are switched on and off via 'epigenetics,' which the chapter defines and explains. In the process we also define and discuss human plasticity, and gain insight into the generationally transient nature of certain forms of human variation. In other words, information in this chapter keeps us questioning the relationship between nature (here, genes) and nurture (the environment, and all that it entails), and demonstrates how nature and nurture work together.

Chapter 4 defines the term culture with reference to its adaptive function. Our capacity for cultural adaptation—once it emerged—has been crucial to humanity's survival. Culture's evolutionary emergence is explored with reference to the synergistic interaction of our ancestors' evolving capacities for motor skills, empathy, cooperation, and symbolic communication (for instance, through the language of ritual). 'Cultural relativism' and 'ethnocentrism' also are defined and discussed as contrasting standpoints for interpreting cultural data.

In Chapter 5 we turn our attention to the spread and settlement of behaviorally modern humankind, made possible by culture's emergence. We examine how human migration from our species' birthplace in Africa, and subsequent movements of various groups to places around the globe, fostered diverse biocultural differences between human populations. We question assumptions about the biology of 'race' and demonstrate how geography in general and geographic clines (geographically specific trait gradients) in particular provide a much better explanation for the biological differences found within our species than the erroneous idea of biological race ever has done or can do. The value of making geographic ancestry-based distinctions between subpopulations in certain limited circumstances is examined.

Central Lessons of Part I

At the end of Part I of the book, readers will be well-versed in systems thinking and able to explain the value of taking a holistic point of view or of adopting a unified approach in exploring biocultural diversity. In addition, they will be able to discuss and explain, using examples, how and why various forms of biocultural diversity came about.

Part I certainly does not ignore the importance of human behavior in the evolution of human variation. However, the overall focus in these early chapters is on geographic or habitat-based adaptations and physical differences amenable to materialist scientific study, such as red blood cell structure, height, and lactase levels. This provides readers a concrete basis for really understanding the concepts presented. In this, Part I fosters in readers a deep grasp of the basics; it does so to ensure success as we study the less typically-examined forms of biocultural diversity that I shall present in Part II and, more so, in Part III.

Anthropology and Complexity

This chapter prepares you to:

- Define biological determinism, discuss its shortcomings, and explain what people hope to accomplish when using it
- Describe anthropology with reference to holism (in contrast to reductionism) and the comparative method
- Define and differentiate ecosystems and complex adaptive systems, paying special attention to the concept of adaptation and the phenomenon of emergence
- Describe the ways in which human groups may be studied as (and as part of larger) complex adaptive systems, focusing on adaptation and emergence

Two of the most important and complicated puzzles humans face can be stated in two quite simple questions: Why are we the way we are? Why are some of us different? These may seem like straightforward questions, but the answers are neither singular nor simple. Begin to brainstorm about flavor preference, sexual orientation, or intellectual capacity, for example, and you will find that what comes to mind are both biological and cultural explanations. Moreover, most groups will split into two camps in regard to the explanations, with one side favoring biology and the other on the side of culture. Which camp is right? Is it our biological nature or is it the fact of having been nurtured in a particular culture that determines how we act, respond, think, move, and feel?

Most people, including most scholars, view biology and culture, or nature and nurture, as different, even opposing forces that work separately or against each other. The opposition between the concepts is, however, both clumsy and futile. They are both part of a system in which multiple and mutlidirectional feedback loops can lead to fantastic and unanticipated biocultural outcomes. This chapter aims to help readers really take hold of this idea by exploring what we know about systems in depth, paying special attention to the concept of adaptation and the phenomenon of emergence in the process. In doing so, we will be implementing anthropology's holistic approach—an approach we learn more about as the chapter unfolds.

Where Did the Nature-Nurture Question Come From?

Biological Versus Cultural Determinism

We are not the first to ask "Nature or nurture?" This question has many historical roots. Looking at our recent past, science writer Matt Ridley has shown that it was incorporated into scientific inquiry in the writing and work of Francis Galton, a productive and well-known nineteenth century geographer, meteorologist, psychologist, and statistician. In 1869, Galton (cousin to Charles Darwin) asserted in his book *Hereditary Genius* that talent is physically inherited: genius runs in the family. Galton was berated by his scholarly peers for ignoring the role of upbringing, so he bolstered his claims with another book, *English Men of Science: Their Nature and Nurture*. As Ridley reports, Galton chose this title purposively: Galton himself explained in the book that it supplied "a convenient jingle of words, for it separates under two distinct heads" the elements that underwrite our personalities (Ridley 2003, 71).

For Galton, the more powerful head by far was nature. By explaining all human characteristics as a result of nature (heredity or biology)—and claiming this 'all' was scientific to boot—Galton discredited the role played by upbringing (nurture, culture). In doing so, he and his followers diverted attention from 'non-biological' causes of human diversity.

We know that **biological determinism**—the argument that biology determines utterly and completely one's capacities and characteristics—has problems. Calling certain arrangements 'natural' is a way to rationalize them. It downplays, for instance, the vast amounts of hard work and sacrifice that records show really are put in by winning athletes such as Venus and Serena Williams, musical superstars such as Ludwig van Beethoven, and intellectual giants such as Albert Einstein. In positioning their achievements as resulting from inborn 'gifts,' rhetoric about them being 'natural' provides a self-serving excuse for the rest of us not to practice, drill, or work out. It diverts us from reaching our full developmental potentials (Shenk 2011).

This does not seem so tragic when we think about it on the level of the individual or when the skills in question are so rarefied. Yes, it is too bad that I am not the excellent soccer player or gold-medal swimmer (and so forth) that I could have been because I avoided really training. However, when applied at the population level, biological determinism that leads people to place false limitations on their potential also can place a damper on progress or productivity, leading to social stagnation. Worse, those in power can use biological determinism strategically, to justify their advantages.

If the genius it takes to fix cars, say, is hereditary in males but not females, then barring girls and women from auto shop classes can be seen as justifiable: to let them in would be futile and waste resources. Similarly, the poor might be kept from being educated on the basis of an argument that they are inherently dim-witted. Certainly there would be no need to examine underfunding in inner-city schools as a possible source of the low achievement rates if they were written off as due to biology. These examples may seem extreme but they represent real situations, present and past

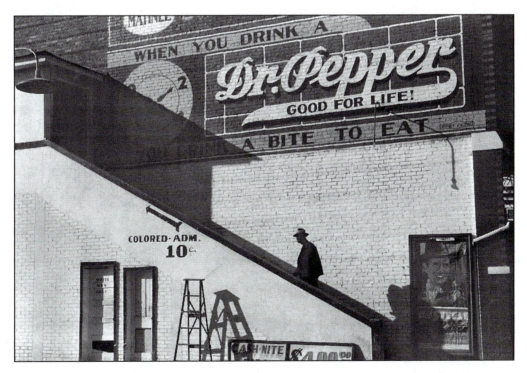

Figure 1.1 "Colored entrance" of a cinema in Belzoni, Mississippi, c. 1939. *Photo by Marion Post Wolcott, Library of Congress Prints and Photographs Division, LC-DIG-ppmsca-12888.*

(see Figure 1.1). Biological determinism has even justified the enslavement of one race by another as part of the natural order.

We know today, of course, that racist thinking has no scientific basis whatsoever (see Chapter 5). The myth of 'giftedness' also has been debunked (Shenk 2011). Yet, biological determinism still haunts us—not only in the way it still is used in justifying inequality but also in its antithesis or counter-form, cultural determinism. **Cultural determinism** argues that culture (rather than biology) determines one's capacities and characteristics. Early twentieth century anthropologists in the United States, appalled by racism, popularized this position. It may have its appeal, but it, too, is one-sided. It, too, exists, in a way, as an outcome of Galton's "convenient jingle of words": with respect to our original questions—Why are we the way we are? Why are some of us different?—Galton and his followers have led us to believe that the answer must be *either* nature *or* nurture, *not both* (see Ridley 2003).

Nature and Nurture Work Together

Recent advances in philosophy and science demonstrate that determinism of either kind is wrong-headed. Rather than asking if nature or nurture is in charge, we should ask how the two forces work *together*, inseparably in a system, to produce the vast and wondrous diversity of human experience. Humans are neither simply biological nor simply cultural; we are both.

To think otherwise is in some ways an artifact of language: separate terms label each of these facets of human existence, suggesting they are mutually exclusive. Any attempt to talk about them as blended ends up resorting to an inserted 'and' or, more efficiently, a silently hyphenated 'biocultural.' The assumption, even here, is that biology plus culture (or nature plus nurture) equals human experience; the two still are separate due to the additive characteristic of any compound phrase or word.

Nevertheless, with some effort, it is possible to get past this language barrier conceptually. Indeed, characterizing the inseparability of biology and culture while exploring human diversity is the focus of this book. To begin to do this, we must first learn a little bit about systems thinking. A central feature of all systems is that, in listing their parts, something more like 'multiplied by' can take the place of 'and' or 'plus': systems are interactive or multiplicative, not additive.

Holistic Systems Thinking

In **systems thinking**, we focus on connectedness. As a point of view, systems thinking highlights relationships and what emerges from them.

Over two thousand years ago, Aristotle noted in *Metaphysics*, "The whole is more than the sum of its parts." Indeed, and to push the mathematical metaphor, it is the *product* of its parts. A system derives from the *relationships* between its parts—not merely from the addition of these parts. A cake is not just a pan filed with eggs, milk, sugar, flour, and butter. It is the thing that emerges through the combination of ingredients plus a cooking process. In other words, *the properties of a system can be neither explained nor determined by examining its parts alone.*

This perspective is called **holism**. It often is contrasted with **reductionism**, the idea that an entire system can be explained by a single aspect of it. For example, explaining organisms such as humans (or ducks, say) as machines reduces them to their moving parts (see Figure 1.2). Doing so might be helpful in certain circumstances, but it misses the mark when

Figure 1.2 Mechanical duck, said to have been built by Jaques de Vaucanson; this imagined interior was drafted by an American observer, 1738–1739, France.

one desires a full understanding of human (or duck) experience. Not only does it fail to bring into consideration other dimensions of the whole human (or duck) being, it also fails to position humans (or ducks) in a larger context—despite the fact that everything is part of something bigger and multidimensional.

Holism is anthropology's hallmark. This in part explains the breadth of the discipline's title: based on the ancient Greek terms 'anthropos' and 'logos,' **anthropology** means, literally, the study of humankind. Anthropology is the study of human systems and human interactions with the systems in which they are embedded (for example, the physical geography of the earth system). Humans affect these larger systems just as they are affected by them—for instance, developing complicated cultural strategies for finding water in arid, drought-prone climates, or evolving a greater genetically-driven capacity for cold tolerance after generations of living in arctic regions.

Ecosystems

One strain of systems thinking is the ecological or **ecosystems** approach, which gained popularity in the United States in the 1960s and 1970s as people began to realize in large numbers that some human actions (such as dumping pollutants in rivers) were harming the environment. Ecosystems include multiple interconnected populations, both **biotic** (living) and **abiotic** (for example, rocks). Most biotic populations—that is, most **species** or breeding groups—are quite specialized, requiring particular resources for survival (for example, a species may eat only a particular bug in a particular stage of its life cycle). Each species therefore generally occupies a particular **niche** in the system, thereby playing a particular role.

An ecosystem's populations are all related to each other through various ties involving energy flows. For instance, squirrels rely on oak trees for acorns (calories), and oak trees rely on squirrels to spread their acorns, thus planting new trees, and to provide, via their dead bodies as well as excrement, fertilizer that sustains the trees. Add in other populations, such as fleas, squirrel-eating wolves, grasses, a river, limestone—even people—and so forth, and we have an ecosystem.

When the ecosystems approach first gained steam, people assumed that an ecosystem had an enduring ideal balance point—a point at which species' population numbers, and the relations between populations (for example, supply and demand), would hold steady even as time passed. A system in balance is in **homeostasis**; it has achieved a 'steady state.' In the original ecosystems model, change is generally seen as bad for balance; it is associated with stress on the system. For instance, if people overhunt squirrels, oak trees could suffer as they lose a source of fertilizer and a mode of seed (acorn) dispersal. A chain reaction is likely to ensue, with one population's demise leading to growth for another, which in turn creates stress for a third, and so on. Something must happen to restore balance, or the entire system may collapse. Viewed from this perspective, change is not good.

Complex Systems

Information and Adaptation

Another approach to systems thinking has developed, and this one sees change as an inherent feature of the system—not a negative disruption to it. This approach stems from complexity science, which emerged with full steam in the context of the late twentieth century's revolution in information technology. (It has since infused ecological science, but for now, to keep things simple, I will first treat complexity theory as a separate entity, which it was when initially invented.)

Some systems are **mechanical systems**, and therefore totally predictable. For example, a thermostat-regulated heating unit will switch on and off when the temperature moves below or above a certain set-point. This system detects one thing—temperature—and responds in one of two pre-determined ways (turning on or off). Human systems, however, are not mechanical. They can evolve adaptively in response to changes in the environment. They bear closer resemblance to **complex adaptive systems**—networks of dynamically interrelated parts between which information can flow and in which "emergent, surprising, creative behavior is a real possibility" (Plsek 2001, 310)—than to mechanical systems. For this reason, the complex adaptive systems framework provides a fruitful source of insight into human existence.

Following this framework, let us assume that systems are driven to try to stay organized, or stay together. Given a change in the environment in which a system is embedded, simple interaction between its parts that strive to keep the system working can lead to something entirely new—to a novel pattern or property: to **emergence**. Many things that we take for granted today, such as our reliance on electronic messaging, or fast food and ready meals, began as emergences. Emergence itself also can lead to unintended consequences, whether on a small or large scale. For example, electronic messaging fosters flexibility in planning that was not possible in the last generation.

Another example is the way the spread of the automobile fostered an unintended increase in traumatic injury because of motor vehicle crashes in addition to changing the landscape (see Figure 1.3). The civilian emergency medical and trauma services systems developed or emerged in turn, largely in response to the rising need for immediate medical attention that motor-vehicle-related injuries created. The health services system was dramatically altered, as was our entire employment and residential structure, by the introduction of the private car, which effectively made suburbs and office parks possible. Other knock-on or secondary (domino, butterfly)

Figure 1.3 Clark Avenue and Clark Avenue Bridge, looking east from West 13th Street, Cleveland, Ohio, 1970. *Photo by Frank J. Aleksandrowic, Environmental Protection Agency, NARA record: 8452210.*

effects are the emergence of the anti-theft device industry, the dangling air freshener industry, the road-map (now, GPS or sat-nav) industry, and travel games for passengers; the list goes on and on. Our social and cultural systems changed radically with the automobile's emergence and those emergences that followed on from it: we adapted.

Importantly, changes like those just described not only illustrate the dynamic, ever-changing nature of complex adaptive systems, they also are **nonlinear**; that is, ramifications do not unfold in a predictable manner but instead as surprises that lead to surprises of their own. Cars led to the drive-in cinemas and drive-through burger stands that took the country by storm in the 1950s and 1960s, which then led to the modern fast-food industry. If someone had told Karl Friedrich Benz, an inventor of the automobile, that he would revolutionize the world's food industry, he is unlikely to have believed it. Nor would Raymond Kroc, founder of McDonald's, have predicted today's obesity and diabetes epidemics when he began his business in the 1950s. The emergence of these things fulfills what is sometimes called The Law of Unintended Consequences. What we do to survive these unintended consequences—to adapt to them—is an open question. The future, from a complex adaptive systems perspective, is indeterminate.

A complex adaptive system is thus one that alters itself, or adapts. Technically speaking, an **adaptation** is a survival-enhancing change in a system brought about in response to some kind of atypical, stress-producing change in the environment. In contrast to a traditionally-conceived ecosystem or a mechanical system, which either breaks down in response to the pressure exerted through such a change or tries to restore itself to some pre-set balance point, a complex adaptive system evolves in tandem with such environmental pressures. It learns from its experiences.

One reason that adaptation can happen in complex adaptive systems (which, as scientists now understand, ecosystems really are) is that the complex adaptive systems model accounts for information flows, not just energy flows. Energy flows are linear and their organization is static—remember the classic ecosystem relationship in which a squirrel relies on an oak tree for nuts, and the tree relies on the squirrel to fertilize the soil. By contrast, information flows can lead agents to instigate changes. For example, humans may change their behavior based on verbal or facial expression feedback from friends, bosses, or teachers; ants may change theirs based on scent and sound signals. Such dynamics characterize complex adaptive systems. Complexity theorists focus their studies on the flow of information and the way in which new processes and structures emerge in response.

Delimiting Systems

Systems thinking is used to examine a variety of human phenomena at numerous levels of analysis. For instance, the state is a system, and so, too, is a city within it, as is the neighborhood, family or household unit, and individual body. Indeed, each of these systems is itself a system within a larger system and home to smaller, internal systems. The class for which you may be reading this book is itself a system that can be viewed as part of the larger university system, and so on.

Any system that is simultaneously a whole unto itself and a part of something larger can be called a **holon**. In theory, holons have boundaries, just as the walls of an apartment or house (or yurt, and so forth) bound the household. The holon concept thus provides a convenient mental means of packaging or ring-fencing sub-systems for discussion or scrutiny, as I did above when I spoke of a system being embedded in an environment.

In reality, however, calling any system out as a self-contained holon is somewhat deceptive. Differentiating a system from its environment rests on a false distinction between the two. Boundaries like walls, or skin, or a village's borders, however real, always are permeable (and even apparently impermeable boundaries are worked around by other connections).

Multidirectional links continue to exist; energy and information continue to flow. Indeed, *if linkages did not persist, changes in the so-called environment enveloping a system could not actually have an impact.*

Delimiting the system under study can help focus discussion, and that is a good thing. However, the contrived nature of the holon is all the more apparent when we take into account that where a holon's boundaries are depends on the observer's standpoint. For instance, investigators who wish to study the mechanics of breathing will focus on the respiratory system, or perhaps the higher-level system of the body. Those interested in studying what is breathed—air quality—may take a neighborhood perspective, or may extend their study to larger systems if they find that pollution is coming from a far-away source.

A college class provides another example—and also shows how getting bogged down in questions about where systems begin and end can distract us from their more important features (dynamic connectivity between the system's parts, adaptation, emergence) just as the nature-nurture debate distracts us from the jointly biological and cultural dimension of human experience. To study a class, we could use the enrollment roster as the concrete, physical delimiter of a given classroom system; alternatively, we could use the classroom's walls—but the class is not just the space, plus the students, plus a teacher, plus course materials: the class is what *emerges* when all these features or system parts interact. The classroom's physical layout, the personalities of participants, and even the time of day we meet are just a few of the many factors that come into play as information and energy flows. If the class is an online class, a key factor is the kind of online learning system or software the school that sponsors it uses, which in turn determines to some degree the shape that learning takes.

When we envision the class as a system, it becomes clear how many things affect learning outcomes: whether students and teachers are timely, respectful, prepared; the temperature of the classroom; the state of the room's furniture and its styling; whether or not someone is eating lunch in the room, or snoring, or sending text messages to a sweetheart (in an online course, some factors might be the quality of the internet connection, where the student is logging in from, and so on). We can see how asking questions in class helps a teacher gauge how well the material has been understood. We can see how learning emerges, for better or worse, from the relationships between all the system parts. We can acknowledge other things that emerge, too: friendships, for example, and the spread of the flu.

Studying Human Systems

The Evolution of Anthropology

Anthropology, we have learned, is by definition interested in human systems, human culture included. During the late nineteenth and early twentieth centuries, when academic anthropology came into its own, anthropologists often made long field trips to far-away places where they could study native non-European cultures. Many spent a great deal of time trying to measure and document human physical diversity and to describe and record native languages. Anthropologists also took pains to create descriptive written accounts of the natives' social structures and cultures—**ethnographic** accounts. Specialists in doing this often also practiced **ethnology**: they made cross-cultural comparisons in order to understand the significance of the cultural variation that they had documented, as well as to test for universality.

A key difference between 'anthropology' and 'ethnology' is rooted in ancient Greek: 'anthropos' means people in general, and 'ethnos' refers to the culture or design for living held to by a particular group of people. It refers to a given group's shared, learned heritage—including the systems this heritage supports ideologically: social, political, economic, spiritual, and so on. In fact, today ethnology is often termed 'sociocultural' or even just 'cultural' anthropology.

Indeed, many anthropologists now focus solely on sociocultural issues. Many identify, equally narrowly, with the physical or biological dimensions of humankind. Other specializations entail linguistic and, in the United States, archaeological anthropology. This four-square partitioning does not fit well with the idea of anthropology as the study of humans in the round—but it does make sense when seen in the context of academia's evolution (and see Chapter 5 regarding four-fold thinking).

From Fission to Fusion

The history of academia as a whole has been one of territory-staking. More often than not this meant carving off a precise aspect of human existence for study. Today, we are noticing problems resulting from the narrow thinking that such subdividing can bring about. The twentieth century seems bound to be remembered as a century of disciplinary fission or fragmentation. With some effort, the twenty-first century could, in contrast, be a time of fusion or cohesion, in which holism pervades not only anthropology but other areas of study as well.

For example, although academic training generally turns out very narrowly specialized researchers, we can self-assemble investigatory teams committed to deriving a holistic picture of a given topic of study. A unified, transdisciplinary, multi-methods approach leads to more comprehensive, realistic, illuminating, and therefore relevant findings. Some large research funding agencies today are in fact demanding integrative, transdisciplinary approaches from their fundees.

The individual researcher, too, can strive for holism by crossing humanly-created academic boundaries. The biological anthropologist interested in population-level differences in rickets vulnerability, for instance, will want to examine not only possible genetic differences, for example in relation to skin pigmentation, but also breastfeeding and dietary practices and preferences; how the built environment, employment patterns, and latitude affect sunlight exposure; cultural views on tan skin; understandings about the risks of sunscreen, and so on. Importantly, these pieces of the puzzle cannot just be studied in parallel. A concerted effort must be taken to explore how they interrelate, and how changes in one part of the picture may lead to or derive from changes in another; a complex adaptive systems perspective should be applied.

Ethnocentrism Is Not Part of a Unified Comparative Method

We got onto the topic of unification when I mentioned the fragmentation that occurred as academia—and anthropology—grew. Beyond the problems associated with the territory-staking, however, any discussion of anthropology's evolution also must address the deeply embedded biases many early practitioners had regarding the peoples studied. For example, many held fast to the theory of progressive cultural evolution, in which other cultures were cast as inferior versions of the ethnologist's. This idea has long since been soundly rejected and labeled 'ethnocentric.'

Ethnocentrism involves putting one's own culture at the center of any interpretation, in contrast to using that culture's norms and values to make interpretations (see Chapter 4). We are ethnocentric when we take it for granted, based on our own cultural habits, that everyone marks birthdays with cake, or knows a particular Bible verse. Frequently, we are ethnocentric because we assume that the way we do things is the best or natural way, or that everything can be understood from our own point of view. Although anthropology warns now that ethnocentrism severely limits our ability to understand other cultures, many early anthropologists went into the field with an unabashed ethnocentric perspective.

Technically, they did not have to: ethnology as such is simply a comparative practice. Ethnologists adopt the **comparative method,** in which cultures under study are compared with each other as a basis for generalizing about humankind; they also may be compared with the cultures of the anthropologists themselves. When anthropologists compare other cultures with their own, they may do so intentionally as ethnologists—but they also may do it unintentionally. The simple act of thinking about and striving to describe other cultures inherently invites—no, requires—self-comparison. However, it does not require ethnocentrism.

It might be presumed that once we know about our tendency to put our own cultures first, we can shut off the ethnocentric bias. Today, many anthropologists recognize that it is not possible to escape entirely one's own cultural framework, but deep reflection on our cultural biases certainly minimizes the limitations they can put on the ethnological endeavor.

Another way to minimize ethnocentrism is to actively participate in the culture under study. Early anthropologists often hired native interpreters to help them in the day-work of data collection and retreated at night to temporary camps or missionary quarters set back from the settlements of the cultures under study. By the 1920s, however, the need for long-term immersion had grown apparent. Rather than living near those under study, anthropologists now lived among them, learning their language, eating their food, and participating in as much of their daily lives as was possible. **Participant-observation** is the name for this approach, and it is the classic method for creating ethnographic accounts.

All ethnographic accounts, old and new, taken together, comprise the **ethnographic record**. The **archaeological record**, by contrast, is made up of all that we know from what people left behind and how those objects were organized. It includes not only the artifacts of their cultural worlds, such as their houses and cooking tools and high and low status objects, but also (in a subset termed the fossil record) their physical remains, such as bones and coprolites (fossilized feces), which provide important biological data.

The ethnographic and archaeological records often are used in combination to form a better picture of the lifeways that are and have been possible and probable for human beings across space and across time, and to reconstruct the evolutionary history of humankind. Throw in contemporary biological studies, and—assuming we turn away from biological determinism, apply systems thinking, and strive to steer clear of ethnocentrism—we can begin to apply the comparative method fruitfully as part of a unified approach to understanding human variation. We can begin to hypothesize productively about the origins, functions, and magnificent range of our biocultural diversity.

Dynamics of Human Biocultural Diversity: A Unified Approach by Elisa J. Sobo, 19–31.

Genetic Adaptation

This chapter prepares you to:

- Explain, with reference to genes and alleles, how traits can be inherited physically and how variation enters a gene pool
- Demonstrate how expressed genetic variation is useful for a group's survival into the next generation
- Understand the basic principles of natural selection at the population level and distinguish between ecological and sexual selection
- Explain and exemplify balanced polymorphism

For the archetypal ecosystem in homeostasis, everything is perfect. Life is balanced and adaptation unnecessary. However, as we now know, we would be mistaken to think that homeostasis entails an actual stand-still; in reality, maintaining a steady state depends upon change, which is constant and inevitable—even when incremental.

Another word for change is **evolution**, and the kind of evolution this chapter deals with is genetic. **Genetic evolution** happens at the level of an entire population. It entails a change in the frequency of a given gene or genes in the **gene pool**—in the entire population's sum total of genes—from generation to generation.

More often than not, human genetic evolution occurs as an adaptive response to an environmental change that we humans in fact have instigated in a process termed, with increasing frequency, **niche construction**. The rise of this phrase, awkward as it may be, reflects a growth in researchers' awareness of species-environment interaction. It refers to all ways in which organisms in a particular group, simply by doing what they do, shape the environmental niche that they occupy just as it shapes them. It acknowledges the ways each species constructs the world (niche) that they, and other species, live in (Odling-Smee, Laland, and Feldman 2003).

A classic example of how niche construction processes can support genetic evolution—and one we delve into at the end of this chapter—entails the adaptive evolution of human resistance to malaria in populations with increased exposure to the disease due to cultural practices fostering a population boom for mosquitoes (which carry the parasite that causes it). In a nutshell, had humans not applied irrigation as they did in the relevant regions, they would not have ever needed increased protection from the disease. Such mutual interactions demonstrate that we are truly biocultural beings. Genes alone do not determine our potential.

Of course, without genes we would not exist. Knowing this prepares us to ask: what is a gene, and how, exactly, does the gene pool change in situations such as the one described above? Ultimately, we seek an understanding of how and to what extent human genetic adaptation depends on as well as fuels biocultural diversity. To achieve this, we begin by building foundational knowledge in genetics and heredity.

Genes and Traits

Put simply, a **gene** is a linear sequence of Deoxyribonucleic acid (**DNA**) that codes for (or, contains the recipe from which the growing body creates or builds) a specific protein or trait. DNA itself comes in long strings, and each string contains many genes—many information storage units. These strings are generally wound up into little wads or structured packages of DNA, called **chromosomes**. In sexually reproducing populations the chromosomes come in paired sets. They are inherited: half come from the male parent and the other half from the female parent. We inherit two copies of each gene, one from mom and one from dad (biologically speaking).

While those new to genetics might associate each gene with a specific bodily trait, such as nose size or singing voice, in reality things aren't that simple. The majority of traits actually are **polygenic**; that is, they develop through the interaction of a collection of numerous genes ('poly' means many or more than one). Most traits emerge from the combination of proteins and biochemical reactions that a set of genes 'codes for' (and see the next chapter).

Genes themselves can be **polymorphic**, meaning they can come in various forms, called **alleles**. For example, the 'mid-digital (mid-finger) hair gene' alleles code for proteins involved in both hair growth and non-growth. The gene entailed in cystic fibrosis, a generally fatal disease in which the lungs fill up with fluid and certain digestive processes do not work well, also has two alleles—one that codes for proteins involved in the disease and one that does not.

Some alleles are **recessive**, or easily dominated, and therefore less frequently expressed; others are **dominant**, making their own instructions take precedence over recessive alleles so that only they are expressed. This phenomenon—this mode of gene expression—was described by the scientist-priest Gregor Johann Mendel even before we knew about genes per se. Examining peas in his monastery's experimental garden, he noticed dominant and recessive inheritance patterns in color and texture.

A **homozygous** gene pair, or person, has two alleles that are the same for a given gene; a **heterozygous** person has two different alleles for the gene in question (remember, we each have two of every gene due to how we are reproduced). A person can be homozygous when it comes to one gene, but heterozygous for another.

Why distinguish whether a given allele has been inherited twice (homozygously) or not (heterozygously)? Recessive alleles must co-occur in an individual for the trait they are implicated in to be expressed. For example, if a child is heterozygous for cystic fibrosis—if he or she received the recessive allele coding for cystic fibrosis from one parent and the dominant allele that does not from the other parent—he or she cannot develop the disease. Alternately, if he or she is homozygous, having *two* recessive alleles, cystic fibrosis is inevitable. If both parents are heterozygous and so carry one copy of the recessive gene, the child has a 25 percent chance of inheriting two recessive copies. To better grasp this, consider Table 2.1. It maps out inheritance using the graphic form known as a Punnett Square, after Reginald Punnett, who famously described it.

Of course, most traits are not **monogenic**—most are not traceable to just one gene; and most are not ruled by simple dominant-recessive relations. Most traits have complex, polygenic origins. On top of that, the cultural contexts in which we live—in which we are nurtured—directly affect gene expression; this has to be taken into account for a full

Table 2.1 Inheritance of a Recessive (r) Trait

		Mother (Rr)	
		R	**r**
Father (Rr)	**R**	**RR**	**Rr**
	r	**Rr**	**rr**

understanding of our genetic dimension. Still, by thinking about clear-cut examples first, we can begin to understand the basic mechanisms of heredity, such as how genes are passed along. Genes are not the full story, as we shall see, but understanding heredity, and its relationship to genetic variation, provides a good first step in deciphering the puzzle of human biocultural diversity.

Evolutionary Change

In the absence of environmental changes and when mating occurs at random, genes will be reproduced in the next generation at a steady rate. This fact can be referred to as Hardy-Weinberg Equilibrium, after the fellows who first wrote about it. When the rate of a gene's reproduction in the next generation shifts, which can happen for various reasons, genetic evolution has occurred.

General evolution can happen in a gene pool due to simple random chance, and this type of change is called **genetic drift**. Genes are shuffled and resorted during sexual reproduction, and sometimes this randomly leads to changes in the frequency of genes. Random events, too, can affect their distribution, such as when a tree in the woods is randomly knocked over (say, by a lightning strike), and the falling tree kills a badger that had been sheltering below it. The shift in the badger gene pool that this unlucky badger's chance removal caused, and the random redistribution of gene frequencies that can happen during reproduction, are examples of what can cause genetic drift.

Drift may not add up to much with a large population, but it can have a noticeable effect when a population is small. If there are only ten badgers, then the loss of one badger is quite a significant reduction; a big shift in the next generation's gene pool can result. If there are thousands of badgers, the untimely loss of one poor badger will not have the same kind of impact.

A second mechanism for changing a gene pool is **gene flow,** which refers to shifts due to migration. This is what has happened to the U.S. gene pool over the past several centuries. Take American eyes, for instance (the Americas comprise two whole continents, but I use the term 'American' in the lay sense to refer to people in the United States). American eyes, generally quite dark prior to the European invasion, became much lighter by 1900, when about half the population had blue eyes. By the 1950s, however, only about one-third of the U.S. population had blue eyes; today, only one in six people does. These changes seem to have been caused by the flow of genes across national borders; other recent cultural changes, for example in mating and marriage patterns, have accelerated the process (Belkin 2006).

Mutation is a third way change gets introduced into a gene pool. When genes are duplicated for reproduction, alterations can occur; a mutation is simply a miscopied gene. Miscopying often happens due to some kind of exposure, for instance to a toxic chemical, radiation, or a particular virus. Culture can give the biological process of genetic mutation a boost: for example, while asbestos is found in nature, the asbestos fiber products that eventually triggered cancers in so many factory and other workers in the twentieth century exist due to cultural invention. Similarly, our chances for exposure to solar radiation (UV-B) have been enhanced by cultural practices that have depleted the earth's protective ozone.

Not all mutations lead to cancer, of course, or allow fatal recessive diseases to persist in a population even when they kill homozygous individuals before they reproduce (pass on their genes themselves). Most mutations that occur during reproduction are silent or neutral, meaning that they have no immediate impact on an offspring's survival and reproductive capacity. Some mutations actually create improvements—changes that increase a group or species' chances for survival and proliferation (remember, a 'species' is an organism type that can reproduce itself, for instance by interbreeding and giving birth to fertile offspring; reproduction of fertile offspring is impossible across species boundaries).

The Role of Natural Selection in Evolution

Natural Selection Acts on Existing, Expressed Traits

This leads us to the key mechanism of genetic evolution: natural selection. **Natural selection** is a process by which the genes for expressed traits that happen to give an organism an adaptive advantage for survival (at least to the point of reproductive success) under given environmental conditions become, in this way, more abundant in the next generation's gene pool. In other words, the survival advantage that these particular traits confer allows their bearers to produce more healthy offspring. More viable offspring—offspring that themselves live on to reproduce—mean that more of the genetic material that gave the first generation a leg up will exist in the next generation of the population's gene pool, and in the next, and so on, until the environment changes yet again. Successful reproduction is the coin of the realm of natural selection.

Charles Darwin, the scientist popularly associated with natural selection, outlined its mechanisms most famously in his 1859 book, *On the Origin of Species*. For Darwin, of course, genes did not exist; he spoke instead of traits themselves. This reminds us that observable, tangible traits—expressed traits—are what natural selection preserves. The advantage provided by expressed traits given a new environmental pressure—that is, **fitness**—is measured in terms of *lower mortality rates*, which means those with that trait are less likely to die in the face of that pressure than those without it, and *higher fertility rates*, which means those with that trait produce a relatively higher number of viable offspring for the next generation. Darwin called this "descent with modification" (Darwin 1859).

The following hypothetical example illustrates this kind of occurrence (it is far too simple to represent reality, but my goal here is to get the principles across; we deal with reality later; and see Chapter 3). Let's say that people in your world only have purple or green skin. For a long time, this variation just existed, much like hair color varies in our world. However, now, due to some change in the environment—in, say, the culture or climate—having green skin provides a selective advantage. Green-skinned people reproduce more.

Whether one has green or purple skin depends (in this just-so story) on only one gene with one dominant and one recessive allele; the one that codes for green pigmentation dominates. You, fortunately, have green skin. Here's the interesting part: whether your greenness is expressed due to you being homozygous for the trait or because you have one copy of the dominant allele makes no difference according to the law of natural selection. It only matters that greenness was in fact expressed or made manifest in your skin. That—not the genes themselves—saved your hide so that you could later proliferate. The genes came along for the ride when you reproduced; in this way, you passed those genes along to the next generation and experienced reproductive success. Note here that reproductive success means not just having more babies, but having fit ones—babies that can survive to reproduce again, as those with green skin will do in this admittedly simplistic scenario.

Natural selection's focus on manifest traits allowed Darwin and others to recognize it despite their lack of knowledge regarding genetics. It also points us to two new vocabulary terms: genotype and phenotype. Your **genotype** is the set of genes you carry; your **phenotype** is the manifestation or measurable expression of your genes 'in the flesh.'

To be clear: your genotype is not the main focus of the selective process. Rather, your phenotype takes center stage. Remember, a person who carries a recessive (purple) allele has no less chance of reproducing than someone homozygous for the dominant (green) one because the heterozygote still is green-skinned. It is only when the recessive allele is expressed, as purple skin, that a person does not live to pass his or her genes on. It is the phenotype, not the genotype, that gives the disadvantage. Moreover, natural selection works upon *existing* phenotypic variation. A trait has to exist already if it is to be selected for.

This points to the fact that, while the general phrase 'genetic evolution' does cover the selective process just described, the more exact title for it is 'genetic adaptation.' The phrase **genetic adaptation** differentiates the particular kind of evolution brought about by natural selection from that brought about by drift, flow, or mutation alone. Those mechanisms just introduce change. In the kind of genetic evolution that occurs with natural selection—genetic adaptation—certain *already existing* phenotypic or expressed traits become situationally more (or less) advantageous in terms of that group's continued survival (see Figure 2.1).

Figure 2.1   Genetic adaptation; occurs when the forces of natural selection lead to increases, in the next generation, of advantageous manifest traits already existing within the population. *Illustration by author.*

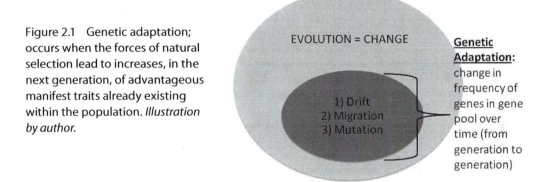

Environmental Pressure and Ecological Selection

Ecological selection—genetic adaptation in response to a change in the environment that makes a given trait advantageous—is a mainstay of natural selection. In fact, and in part for this reason, the phrase natural selection often is shorthand for it.

Let's imagine that a new **pathogen**, or germ, enters and thus becomes part of our environment. Imagine that it is a lethal disease-causing germ. People who happen to be endowed with some pre-existing genetic trait that makes them less susceptible (or more resistant) to the pathogen will have a better chance of survival than will those who do not have the protective gene. Naturally, these protected people will have more children than their dead compatriots, leading to an upward shift in the frequency of the adaptive gene in the next generation. Should the same pathogen sweep through again during the next generation's lifetime, the death toll will be lower: a higher proportion of the population will be resistant.

The imaginary pathogen is but one example of an **environmental pressure**. This category includes anything in the environment that reduces a population's ability to function or puts a damper on its

Figure 2.2
Varieties of
beans. *Photo
by Debora
Cartagena,
Centers for
Disease Control
and Prevention.*

potential, for instance by increasing **morbidity** or sickness, and **mortality** or death. Disease, deprivation (through drought, famine, and so forth), disaster (flooding, earthquake), and an increase in predators are all environmental pressures.

To better understand how ecological selection works, we might go right now to collect some dry beans (or jelly beans or buttons)—their colors do not matter, so long as there are a variety of colors, and so long as some match the color of the table or desk to be used for this experiment (see Figure 2.2). Variety in color represents pre-existing variation in this population's gene pool. Once the beans have been collected, let us assume that the bean population has lived in peace without a predator for many generations.

Once the beans are on the desk, it becomes clear that those beans that match the desk are to some degree camouflaged. Right now, that provides no advantage; but what if a predator did come onto the scene? What if the local bean-eating bird population got so large that some birds moved into the room? This could create environmental pressure on the room's bean population. Bean color could now become very important. If the birds can see some colors of beans better than others, those beans will be eaten faster. The beans more likely to be eaten would therefore be less likely to contribute their genes to the next generation. Beans of their color would become exceedingly rare as selective pressure (hungry birds) removed them. By contrast, beans the color of the desk would have a selective advantage because they can better go unseen; their color would be selected for.

Readers will know by now that new colors are not created in response to hungry birds; the colors pre-exist. The bean population's total color scheme changes as some of those with a disadvantageous color are eaten off. It is true that a bean in a new color may emerge through a random mutation, but random mutations occur independent of environmental pressures. A randomly introduced color may be selected for after it arrives; random mutations are not a responsive development (regarding responsive genotypic, phenotypic, and epigenetic change, see Chapter 3).

Other Kinds of Selection

Non-random Mating (Sexual Selection)

Natural selection via environmental pressure works like a filter on variation that is already there in the gene pool due to mutation, drift, or flow. Ecological selection is natural selection's workhorse. It is, however, aided somewhat by another mechanism: non-random mating. **Non-random mating** is exactly what it sounds like it is: mating and, subsequently, reproduction limited by particular factors determined, whether consciously or not, by the population in question.

Among human populations non-random mating occurs when kinship or class-related rules or ethnic boundaries limit who can build a family with whom. We see non-random mating, for instance, when a person of one ethnicity or religion prefers a person of that ethnicity or religion to all others as a mate. We also see it when certain physical features are deemed attractive, perhaps because they index good health or some other culturally desirable characteristic (for instance, financial standing, moral standards, loyalty to the group; see Chapter 12).

Non-random mating happens even among non-human animals, although only humans would conceptualize it in the terms we are using in this discussion. Female peacocks (pea hens) are more likely to bear offspring with male peacocks that have bigger, brighter tails. Female lions are more likely to bear offspring with male lions that have fuller, glossier manes. The trait that attracts one to another is often termed an **index trait**; it is not important in itself, but rather for what it indexes or symbolizes (in this case, better health, including healthier sperm, which helps insure the viability of the next generation). Note, too, that scent can be as important as visual indicators.

Another label for non-random mating is **sexual selection**. This term highlights the contrast between this kind of selection, which organisms themselves have a hand in, and ecological selection, or the kind brought on by extra-species or, in our case, extra-human factors—for instance, drought or a predator's evolved ability to run faster or bite harder or even outwit its prey. It is important to remember that sexual selection refers to non-random mating, not to sexual reproduction itself; the latter is just a way of making more individual organisms—of reproducing.

Artificial Selection

Speaking of reproduction, **artificial selection** entails the purposeful breeding, by humans, of plants or animals to produce in their offspring particular traits, such as bigger corn kernels, sweeter oranges, faster race horses, or better-nosed hunting dogs. Darwin himself brought observations of human plant and animal breeders' manipulations to bear in his formulation of natural selection. He figured that if "feeble man" could invoke generational changes in domesticated crops, stock, and pets through manipulative breeding, there might be "no limit to the amount of change, to the beauty and infinite complexity of the coadaptations between all organic beings, one with another and with their physical conditions of life, which may be effected in the long course of time by nature's power of selection" (Darwin 1859, 109).

Because human interference was—and is—so explicit in plant and animal breeding, artificial selection provides a fine contrast to natural selection, and thereby to ecological and sexual selection (see Figure 2.3). It is worth considering, however, the ways in which this dualistic mode of classification can, much like the mind-body scheme that pits culture and biology in contrast to one another rather than as two sides of the same coin, lead understanding astray.

Figure 2.3 Natural (ecological, sexual) and Artificial Selection. *Illustration by author.*

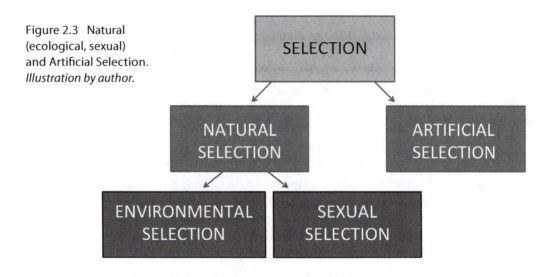

Humans are part of nature, so to call humanly orchestrated selection artificial is illogical. Moreover, so-called natural selection also can have a human handprint. For instance, humans can bring on environmental pressures like those implicated in ecological selection through various means, such as when certain agricultural practices led to conditions that fostered the natural selection of the sickle cell trait as protection against malaria (more of which later). Further, the line dividing artificial selection from non-random mating based on conscious criteria is a fine one, particularly when arranged marriage is considered.

Problems with commonly accepted terminology and related assumptions notwithstanding, if we are to engage in intelligent discussion regarding human diversity, we do need a vocabulary from which to start. We do need to know what is taken as common knowledge. Concurrently, we need not accept common knowledge just because it is common; classification schemes can and often do fragment processes and outcomes that in real life exist holistically. Only if we keep this in mind—only if we refuse to assume that nature is somehow separate from human existence and instead continue to look for connections between (or reunify) dichotomous constructs such as nature-biology and nurture-culture— will our exploration of human diversity be productive.

Co-evolution and the Importance of Within-Population Variation

Remember Darwin's reference to "coadaptations between all organic beings, one with another" (Darwin 1859, 109)? This reminds us that adaptation does not occur in a vacuum; all populations are continually and conjunctionally adapting—or failing to do so.

Whenever a gene pool with some variation meets an environmental pressure, natural selective processes go to work, and genes that provide some kind of advantage will occur more frequently in the next generation. If there is little or no variation in the gene pool, such that no subset of individuals has an in-built genetic advantage, the environmental pressure may in fact extinguish the entire population. This could happen, for instance, if a new and lethal pathogen arrives on the scene, or if an existing pathogen evolves a fatal strain. It could happen through climate change. A population in which genetic variation is not maintained is at a distinct disadvantage in a world that is constantly producing new environmental pressures.

This idea also is known as the **Red Queen Hypothesis** after a character in Lewis Carroll's *Through the Looking-Glass* (1871), the sequel to *Alice's Adventures in Wonderland* (1865). At one point in the story, in a garden of talking flowers, Alice meets the Red Queen. They begin to walk. Soon they find themselves trotting along—and getting nowhere, to Alice's but

Figure 2.4 Alice and the Red Queen running in place. *Illustration by John Tenniel for Chapter 2 of Through the Looking-Glass by Lewis Carroll, 1871.*

not the Red Queen's surprise. The world around them has begun moving. In the Red Queen's world, learns Alice, you run just "to keep in the same place" (see Figure 2.4). Of course, neither the Red Queen nor Alice remains in the same spot literally, because the spot itself differs over time, as do they: the illusion of standing still—represented in the construct homeostasis—rests in fact on constant change. A gene pool in an evolving environment must produce variation from generation to generation just to survive the changes that surround it.

Happily for us humans, like so many living creatures, we reproduce sexually. In sexual reproduction genes are shuffled around, with half coming from one parent and half from the other. This means that every offspring can receive a slightly different combination of genes. In other words, sexual reproduction ensures variation. By contrast, some gene pools do not reproduce sexually and do not produce variation; examples include many single-celled organisms that reduplicate by splitting themselves in two.

If a population reproduces asexually, though **cloning**, each generation is, at least in theory, an exact duplicate of its precursor. This can leave a population vulnerable to extinction because there is generally no subset that might exhibit a genetic advantage. Scientists have learned recently that clonally reproduced species do have a few means of varying themselves. For one thing, mutations can occur as genes are split. For another, external contingencies, such as a temperature shift, can invoke particular adaptive changes through means that scientists are only beginning to explore (see Chapter 3). Still, with clones, what we mostly have is a whole lot of faithful facsimiles.

Although cloning does not completely condemn a group to consist of absolutely identical beings, sexual reproduction *ensures* variation. Even without mutation and flow, simply through the gene rearrangements and re-couplings that occur as part of the sexual reproduction process, there will always be some variation among individuals that can help to insure against catastrophic extinction in the face of a potentially fatal environmental pressure. In this context, human diversity is good.

Natural selection is always at work. However, human evolution in action is hard to detect because each human generation lives such a long time and because our species is now so easily able to move from place to place and intermingle (causing additional change via gene flow, not genetic adaptation). By contrast, it is easy to see natural selection at work in creatures with quicker generational turnover, such as fruit flies or certain germs. We also see it in pathogens that develop a **resistance** to antibiotics: a dose may kill most of a given germ the first time you take it, but those germs genetically endowed to resist the drug then produce many more offspring than those killed off before reproducing, and soon the antibiotic-resistant germs (with the gene that confers resistance) will predominate in the gene pool of that germ.

Examples of Natural Selection

Another creature that has lent itself well to the study of natural selection is the peppered moth, *Biston betularia*. This moth is found in many places but most famously in the forests of England. Two hundred years ago, the majority of peppered moths living in and around Manchester were lightly colored, which camouflaged them when they alit on the light colored trees and lichen in the area. With the onset of the Industrial Revolution, however, the trees (like other bits of England) became dark with soot; much of the light-colored lichen died off. The lighter moths were no longer hidden from sight.

Meanwhile, their darker kin now could not be seen as well as before, and so those moths gained the advantage. Very swiftly, the population as a whole became darker. This process is termed **industrial melanism**, after the industrial processes that darkened the environment and the melanic pigment expressed in the darker moths. Safer now than ever from the birds that otherwise would have preyed on them, the darker moths flourished. The lighter moths, now more easily seen, were pecked off in larger numbers for the birds' lunches. Natural selection worked as a filter on the existing color variation. Had there been no genetic color variation, the entire population might have been wiped out (see Figure 2.5).

This kind of genetic adaptation is termed **microevoluton,** because it occurs on a relatively minor scale (the kind of genetic adaptation that brings about whole new species is **macroevolution**). Examples of microevolution via natural selection in people may be hard to find at first, but they do abound.

Figure 2.5 Peppered moths. *Photos by Olaf Leillinger, 2006. CC-BY-SA-2.5.*

One classic example is the evolution of **lactose tolerance**: the ability to digest milk sugars (lactose) until late in life. For most of human history, the ability to digest milk disappeared in late childhood as the genes responsible for creating lactase, the enzyme that breaks down lactose, turned off. In typical foraging or hunter-gatherer populations (the norm for most of time) lactase is unnecessary after breastfeeding is done; milk from animals is not a food source, and so someone who cannot digest milk is not disadvantaged in comparison to someone who can.

In populations that began to keep milk-bearing animals, however, those who could digest animal milk were generally better fed than others—and had more offspring as a side effect. Eventually, those groups evolved, genetically, so that the majority of their people could digest milk until later in life. This evolution occurred in Northern Europeans as well as in a number of cattle-herding groups in North Africa. The rest of the world's population had no selective need for lactose tolerance and so lactose tolerance is still the exception, not the norm.

Likewise, some scientists think that the ability to digest starches easily was selected for when agriculture was invented and we began to rely more on grains (see Chapter 7). Amylase, an enzyme that breaks down starch, is secreted in our saliva. People who come from agrarian stock have more copies of the gene that codes for this enzyme than do those who come from hunting or fishing populations. Like the ability to digest milk, the ability to digest lots of starch quickly and easily may be a relatively recent adaptation (Wade 2010).

Balanced Polymorphism

As the digestive examples suggest, adaptive traits can be around in a portion of a population for a long time without consequence. They persist here and there because they do no harm. However, once the environment changes (for example, once people started raising dairy cows or grain) certain traits may confer a sudden advantage, and so their frequency quickly escalates. The example of the peppered moth illustrates this process. Other changes in the physical environment, including those related to migration, for instance, from a hot to a cold place or a bright one to a dim one, also can instigate genetic adaptation.

Sometimes, a potentially harmful gene variant can even be selected for. This is what has happened in some regions where malaria is a problem, including parts of Africa and Asia. **Malaria** is a disease caused by parasites belonging to the genus *Plasmodium* and carried by the *Anopholes* mosquito. However, malaria is more than a 'naturally occurring' disease threat; it exists as an **anthropogenic** or humanly generated condition.

Changes in human agricultural and settlement patterns (that is, in human niche construction) have fostered population growth among malaria-spreading mosquitoes and so malaria parasites. For instance, deforestation—clearing land of trees and undergrowth for farming—harms certain bat and bird populations. With fewer of these predators around, more mosquitoes survive to reproduce. Deforestation also reduces plant life that preserves topsoil, so more standing water is available—and that is where mosquitoes reproduce. Irrigation ditches, too, provide lots of lovely standing water, further bolstering the *Anopholes* population boom (see Figure 2.6).

The disease caused by the *Plasmodium* parasites that the *Anopholes* mosquitoes can carry entails cyclical high fevers, headaches, and often death in humans. Indeed, malaria is one of the top ten causes of death worldwide, killing one to three million people annually, mostly children.

In a nutshell, malaria parasites enter the body via the mosquito bite, spend some time in the liver, and then invade the red blood cells. Red blood cells, which carry oxygen throughout the body in their **hemoglobin**

Figure 2.6   Standing water created through irrigation practices. *Photo by Lynn Betts, USDA Natural Resources Conservation Service.*

molecules, are normally round or doughnut shaped. To the malaria parasite, healthy red blood cells are lovely, fat, round nests. They live there and reproduce, damaging and bursting the cells, after which they invade fresh red blood cells as part of an amplification process. They always need healthy red blood cells to proliferate.

There is genetic variation in our red blood cells. One of the alleles for their hemoglobin molecules causes the structure of the cell to be fragile, prone to what my dad would call 'collaption': such cells often take on a crescent or sickled shape as they cave in (see Figure 2.7). These frail cells are not appealing to the malaria parasite: they do not provide good nests. So, in an interesting twist, the more sickle-prone cells you have, the less vulnerable you are to malaria.

Given this scenario, one might expect the sickle-cell allele to have been selected for and the typical version to have been selected against over the generations in areas where malaria is endemic or a constant threat, such as where humans have introduced agricultural and other practices supportive of *Anopholes* breeding. Not so, however. Why? Because too many sickle-prone cells will also kill a person. Sickled cells tend to get stuck in capillaries (the body's smallest blood vessels),

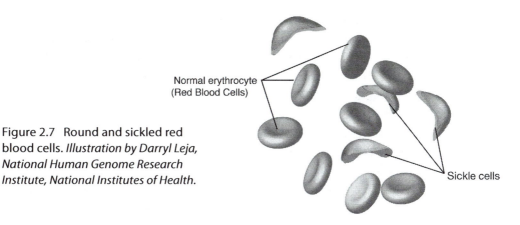

Normal erythrocyte
(Red Blood Cells)

Sickle cells

Figure 2.7   Round and sickled red blood cells. *Illustration by Darryl Leja, National Human Genome Research Institute, National Institutes of Health.*

jamming up on each other like logs in a narrow stream. The knock-on effect of the sickling process is a disease known as **sickle cell anemia**, or sickle cell disease, which entails, among other things, crisis periods in which individuals suffer great pain, as well as early death. So we are caught on the horns of a dilemma: the person with 100 percent round red blood cells is a good host for malaria and more likely to die from it; the person with 100 percent sickling cells will not die from malaria— but he or she will die from sickle cell anemia.

Sickle cell disease is sometimes called a recessive disease—one contracted when an individual inherits two copies of (is homozygous for) the sickle cell variant of the hemoglobin gene. In truth, however, sickle cell disease is not fully recessive; the heterozygote who carries the trait in just one allele still can show symptoms because some of his or her red blood cells, but not all, will sickle. Importantly, though, he or she is likely to be healthy enough to reproduce, thus contributing genes to the next generation's gene pool. Moreover—and this is why the heterozygous condition persists—when in a heterozygous pair, the sickle cell allele provides just enough protection from malaria to maintain itself.

Given the anthropogenic threat of malaria—that is, in concert with a culturally-created situation—natural selection promotes a balance between the round and sickling variations of the hemoglobin allele. The fancy label for this situation is **balanced polymorphism**. In balanced polymorphism, the pressure for one form of the allele is balanced or offset by the pressure against that form. An allele that itself is deleterious or harmful can be kept in a population's gene pool in this fashion because the hazard it entails is offset by the way in which it protects the organism from another problem.

Where malaria has been tamed through cultural practices and ceases to pose a risk to human health, the sickle cell allele will offer no benefit and so could be cleared from the gene pool eventually. Such an adaptation, in which human behavior begets a change in the status of malaria that in turn begets a change in the frequency of the sickle-cell allele, demonstrates clearly the interlinked nature of biology and culture.

We will close this chapter on the malaria-sickle cell example, which amply illustrates important concepts in genetics and in regard to heredity, and which, moreover, makes manifest the value of the adaptive processes that produced diversity in the distribution of the sickle cell trait to begin with. If populations exposed by their own agricultural and settlement patterns to malaria hadn't in turn evolved, via natural—more specifically, ecological—selection, to feature a high frequency of sickle cell alleles, these populations might not have persisted. The synergistic intersection of natural and cultural forces within the human body underwrites the global distribution of biocultural diversity.

3

Developmental Adjustment and Epigenetic Change

This chapter prepares you to:

- Define developmental adjustment and distinguish it from natural selection
- Identify, define, and distinguish the genome and the epigenome; discuss both in relation to the differences that exist between species as well as between various populations of the same species
- Describe how developmental adjustment and epigenetic events affect gene expression, or help organisms translate a genotype to a phenotype
- Delineate the paradigm shift model of scientific revolution; show how it applies in regard to epigenetics

In the preceding chapter we learned how human variation supports the survival of human populations over time. We distinguished the genetic makeup of the individual—the geneotype—from the physical manifestation of genes in the flesh—the phenotype. Although we often think of the most visible phenotypic features—such as hair, skin, and eye color—when considering our genes' concrete expression, many phenotypic features are invisible. Take, for instance, how much hemoglobin (oxygen-binding molecules) a given person's body can produce, or the average shape of the red blood cells that are home to the hemoglobin. One's phenotype is manifest both outside and inside.

How do genotypes become phenotypes? We know that most genes, in effect, contain information in a code that can be translated, by other cellular functions, into a particular protein, and we know that proteins are physical building blocks. While such basic information sets the stage for understanding human diversity, getting from genotype to phenotype is not as simple as building bodies up from a series of genetically pre-determined proteins. Our incarnation is no simple matter, and a simplistic vision of DNA as a blueprint carved in stone will no longer suffice.

We have already discussed the expression of dominant genes, comparing recessive disease 'carrier' heterozygotes with unaffected, non-carrying homozygous individuals: their genotypes differ (for example, Rr vs. RR) but their phenotypes are the same. Both remain healthy, or anyhow do not suffer from the recessive disease. However, the dominant-recessive situation, in which certain genetic instructions (those of the recessive allele) remain suppressed or unexpressed, actually is quite rare. Understanding dominance and recession is a first step to understanding genetic adaptation, but it should not divert us from the fact that many if not most traits are polygenic (based in many genes) and many alleles are co-expressive (that is, members of the inherited pair both affect the phenotypic manifestation of the relevant trait).

Extra-genetic factors also come into play, affecting phenotypic expression, and in fact acknowledging this has led to a real shift in scientific thinking about how we turn into who we turn into, and how that process can differ across populations. Figuring out just what the relevant extra-genetic factors are and how they make their impact is the point of studying what we will for now term 'developmental adjustment.'

Developmental Adjustment

A **developmental** process is one that occurs while an individual organism is growing or maturing. **Developmental adjustment** refers to a long-term change in an organism's growth or biochemical processes during development that occurs in response to environmental conditions. Some alternate terms for this are 'developmental adaptation' and 'developmental acclimatization.' **Acclimatization** (or **acclimation**) can refer to shorter-term, reversible self-protective changes, such as the seasonal tan that some summertime beachgoers develop, so for clarity I avoid using acclimatization in a developmental context. Likewise, I prefer developmental 'adjustment' to 'adaptation' because, although a developmental adjustment is adaptive in a general sense, adopting the term 'adjustment' instead signals key differences between the two processes. Genetic adaptations are inheritable and occur at the population level, affecting the next generation's gene pool. Developmental adjustments happen in an individual's lifetime rather than as a generation-to-generation shift, and they are not directly passed along to any offspring (see Table 3.1).

Table 3.1 Genetic Adaptation versus Developmental Adjustment

	Genetic Adaptation	Developmental Adjustment
Cause / Mechanism	Natural selection: Lower mortality and higher fertility rates among subpopulation with advantageous trait	Pre-natal and childhood exposure: Physical plasticity responsively optimized during organism's development
When realized	Next generation	This generation
Directly heritable	Yes	No

For example, a boy who grows up in a location at a high altitude, such as in the Peruvian Andes, will probably develop a larger chest and greater lung capacity (and more red blood cells and certain other traits) than he would have had he been raised at sea level. His body responds to the lower concentration of oxygen at high altitudes, as bodies will do, by effectively growing bigger lungs and so forth. If, however, when he is older, our boy moves to a low altitude and has children, they will not automatically have large or so-called barrel chests just because he (their father) does. Their chests will develop at an altitude-appropriate rate. Similarly, two identical twins raised apart, one in Tibet or Peru at a high altitude and one in, say, San Diego, which is more or less at sea level, will end up with different chest phenotypes despite their genotypic equality.

Now certainly the **plasticity** (malleability, developmental range) underlying the potential for developmental adjustment is itself genetically determined. Moreover, our genes delimit a certain point past which growth is not possible. Think about the distribution of height that we would see if we lined up all the students in a class. While there will be an average, there also is an upper limit. Similarly, there is a range to chest growth at high altitudes. In a given high-altitude population, some individuals will be born with DNA that supports more expansive chest growth than others. These individuals may be better fitted to the high-altitude environment through their genetic endowment, and so they may survive to have more children than low-growth-potential individuals will. They will pass along their high-growth-potential genes at a greater frequency, thereby altering the gene pool. Yet the accomplishment of extra chest growth, however extensive, is not itself passed on to progeny. That is because it depends on developmental circumstances, not genes per se.

Beyond chest growth at high altitudes, there are plenty of other examples of developmental adjustment. For example, take height. Barry Bogin (1998), who has studied height for many years, reports that while in 1850 Americans were the tallest people in the world (with men averaging 5'6" in height and women 5'1") they now are in third place, with men at an average of 5'8" and women 5'2". The Dutch, whose men were among the shortest in Europe in the 1850s, are now number one, with their men averaging a height of 5'10" and women about 5'5".

Bogin asks, "So what happened? Did all the short Dutch sail over to the United States? Did the Dutch in Europe get an infusion of 'tall genes?' Neither" is his answer (1998, 41). The change had more to do with nurture (the culturally-created environment) than nature (genes). Living standards improved in both the United States and the Netherlands, but they improved more, on average, for the Dutch, and in the process height in the Netherlands increased more than height in the United States. Height is, for sure, due in part to genetics, but environmental factors also affect developmental achievement, for example leading our bodies to grow parts larger or smaller as needed. For instance, an undernourished child's body growth will be reduced by circumstances: when raw materials found in food are in short supply, the individual cannot build as many cells as he or she otherwise would. This results in shorter adult stature—but also better chances for survival than he or she would have if scarce food resources were 'wasted' on growing his or her body tall.

Bogin's data also show the reverse. Maya refugees to the United States during the 1980s found security in their new country, including abundant food and treated drinking water. After just a short while living here, their children were bigger: first generation immigrants were 2.2 inches taller than their mothers and fathers. Similar but less dramatic increases were shown for immigrants from Europe in the later 1800s and early 1900s. In fact, anthropologist Franz Boas used anthropometric (body size) data reflecting this plasticity to argue against the racist version of biological determinism prominent at that time.

Epigenetics

In the past, it was typical to paint the contrast between developmental adjustment and genetic adaptation in stark black-and-white terms. The two were distinct, and overlap was denied if not muted. However, in recent years, scientists have seen that the overlap is even more extensive than the examples just given suggest. Scientific advances reveal that assumptions about DNA's directive role and its role in separating species have been vastly too simplistic.

Different Species Are Not So Different

We have actually known about problems with late twentieth century assumptions about DNA for quite some time. For example, in the 1970s scientists discovered that near to 99 percent of the DNA in humans and chimpanzees is identical. Despite this fact (or perhaps because of it), in 1990 the scientific community began a quest to map every single gene in the human **genome**—the entire stock of genetic information carried by the species. Thirteen years later, the project was complete. Concurrently, other scientists were studying the genomes from other species. As a result, the assumption that more genes mean more complexity was debunked (Ridley 2003, 24–29). Moreover, while recent revisions to original estimates reveal that we have 20–25,000 genes, *C. elegans,* a tiny and essentially brainless but much studied worm, has about 20,000; the *Drosophila* fruit fly does, too (*Economist* 2007). In addition to being similar in number, our genes are not far different in kind from those of other mammals. How, then, does each species attain such different phenotypic results?

Figure 3.1 Side view of human skull (M) and gorilla skull (G). Alfred George Fysh Machin, *The Ascent of Man by means of Natural Selection* (1925): Longmans Greene and Co, London.

Ridley (2003) helps explain by pointing to the difference between *David Copperfield* by Charles Dickens and J.D. Salinger's *Catcher in the Rye*. Using mostly the same words, Dickens and Salinger have produced two very different books. The difference, says Ridley, lies in the pattern and order of word use. Likewise, gorillas and human beings have and use many of the same genes, but each in a different pattern and order. So, for example, gorilla heads and human heads are visually distinctive because gorillas grow their jaws for a longer time and their craniums for a shorter time than human beings do. Their jaws protrude more than ours do and our craniums are higher (see Figure 3.1). As Ridley says, "The difference is all timing" (34).

Consider human and gorilla skeletal structures as a whole now, and, for added effect, let's throw in the skeleton of a zebra, say. Each species has not only a skull, whose constituent parts grow more, or less, depending on genetic programming, but also a rib cage, four appendages (we call ours arms and legs), a backbone, hips, and so on. The pieces are in some ways quite similar; the difference has to do with where a bone curved in the growth pattern, how long it was allowed to grow, and so forth. The bones in the human tail, for example, stop growing pretty early in comparison to the tail of the zebra. Now bring into our comparison chickens, say, and pythons. Each has thoracic (chest) vertebrae; but the number of vertebrae—the extent to which the thorax is directed or allowed to grow—differs so that in chickens it is limited and in pythons it is extensive; that is, a python's thorax extends all along its snakey body. The illustrations in Figure 3.2 help make the point clear: the same kinds of limb bones follow very different growth patterns depending on which creature owns them.

All Genes Are Not Alike

Bone growth timing (and the timing for growth of other bodily components) is regulated in many ways. Some gene regulation is, literally, genetic; that is, instead of or in addition to coding for proteins, some genes act as regulators, signaling a production schedule to the dedicated protein-coders. Regulators tell protein coders to turn on or off, for example.

Broad acceptance of this fact has been a long time coming. The original definition of DNA as a blueprint directed attention to its protein-coding bits and away from genes or parts of genes that could not be associated directly with proteins. The latter often were cast as mere filler. However, today's scientists recognize gene regulation as where the real action lies. Scientists also now know that, in addition to receiving signals from other genes, protein coding genes receive work (and stop-work) orders from beyond the thread of DNA that contains them.

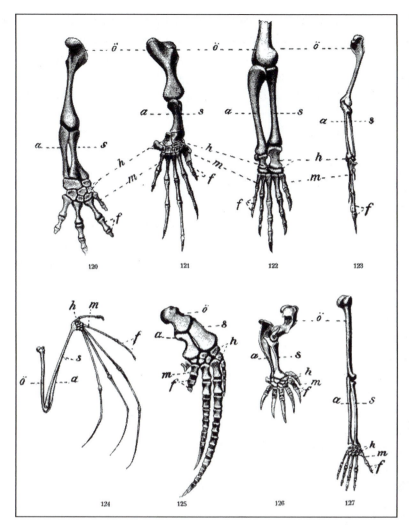

Figure 3.2
Comparative study
of the skeleton of
the arm. Wilhelm
Leche, 1909.

Epigenetic Gene Regulation

The term **epigenetics** was first promoted by Conrad Waddington in the early 1940s. He defined it as the study of interactions between genes and their environment that "bring the phenotype into being" (1942). By the late 1950s, he was using a landscape metaphor to help others understand the systems perspective he promoted. Although the multidimensional "epigenetic landscape" Waddington described was microscopic, his vision resonates exceedingly well otherwise with what we learned about developmental adjustment and how the environment (given or culturally created) affects the phenotypic expression of a genotype.

A newer definition—one that drills down even deeper into the mechanisms involved—defines epigenetics as the domain of science focused on the **epigenome**: the layer of biochemical interactions surrounding our genes. Epigenetic biochemical events turn genes on and off, or otherwise affect their expression. When we think of this layer of biochemical interactions as enveloping, layered upon, or wrapped around the DNA, and recall that 'epi' is Greek for upon or above (the top layer of skin is the 'epidermis'), the term 'epigenetics' makes good sense. It must be said that its definition in the 1940s, when we knew so little about genes themselves, was not so narrowly focused. Nevertheless, right now, it is in the epigenome that many scientists think what really makes us human will be found.

Genetic expression is regulated when some genes create **epigenetic events** that turn other genes on or off. It also comes into play through **RNA** (ribonucleic acid), DNA's self-generated biochemical helpmate. RNA once was thought to simply fetch and deliver the proteins that DNA called for, so that bodily parts could be constructed according to plans contained in the DNA. However, RNA can actually take a more active role in the process: some kinds of RNA can regulate gene expression and mediate the translation of genotype to phenotype versus simply providing fetch-and-carry services. Indeed, RNA seems to hold the key to many notable species-to-species differences.

For instance, when Ronald Plasterk's lab compared particular kinds of RNA expressed in the human brain with those of chimpanzees, it turned out that 8 percent were unique to human beings. This was significantly more than would have been predicted based on chance or the 1–2 percent difference between chimp and human genomes to begin with. As noted in news coverage garnered by this finding, we are only now discovering the extent of human RNA's uniqueness and centrality—in part because until recently we simply did not think to question accepted knowledge about RNA passivity (*Economist* 2007).

Humans actually have about twice as many RNA forms as DNA forms (genes), and their RNA is very busy. Indeed, human cells host a huge diversity of RNA types, doing a vast diversity of functions—including regulating our protein-coding genes so that collecting those proteins and building with them either starts, stops, or continues. The duration for which a gene gets to stay switched on—the degree to which a gene gets expressed—seems essential in differentiating humans from other creatures and, moreover, humans from humans. As a review of the situation notes, "In the complexity stakes, it is not how many protein-coding genes you have, but how you regulate them that counts" (*Economist* 2007). In light of this, it may be more helpful to view DNA as an RNA factory versus a set-in-stone blueprint for human development, and to look to epigenomic processes for information on what makes humans human and diverse.

Environmentally Induced Epigenetic Events

It may seem unusual for science to reach backward, but Waddington's early definition of epigenetics as gene-environment interaction that leads to a phenotype's emergence and his landscape metaphor were quite perceptive. Beyond RNA, and the genes or DNA segments that themselves serve as regulators, there is a third source for epigenetic events: biochemical components arising, in humans, for example, not from our own pre-programmed design but rather from, for instance, food we eat or toxins we get exposed to. The environments in which we live make a direct difference to our genes—not just in the next generation but now.

For example, as Dana Dolinoy has explained, mice prenatally exposed to a particular toxin exuded by certain plastics suffer over-expression of a particular gene called agouti, leading to obesity and a propensity for cancer. The agouti gene will not shut off after exposure to the toxin (Nova ScienceNOW 2007). Some human cancers seem to work this way also, and great promise lies in the possibility of interfering therapeutically with the epigenome's on-off switching mechanisms. Scientists hope the switches can eventually be manipulated by doctors—or through changes in diet and lifestyle that in turn can alter a person's exposure-linked risks.

Other examples of the environment's impact upon genetic expression abound. Take, for instance, so-called animal instincts. It is said that a duckling instinctually follows its mother after being hatched. Its instinct, however, if we can still call it that, really is to follow the first moving body it sees during a certain window of time in its early life. In most instances, the body the baby duck sees will be its mother; her image will be imprinted, so to speak, in its mind. However, if the duckling first sees a different moving object, such as a person, it will assume that that person is the one to follow (see Figure 3.3). There is another interesting thing about ducklings—they are 'naturally' attracted to the call of other ducks. However, it turns out that this is only because such calls are the first calls that they hear: they hear their own calls while they are still in their eggs. If their vocal cords are cut before they hatch, they will not have a preference. Their 'instincts' thus depend quite clearly on environmental cues (Ridley 2003, 152).

People, too, are affected by early exposures. For instance, infants learn flavor preferences through breast milk and even prenatal exposures (Mennella, Jagnow et al. 2001). Children who live in environments necessitating they run often and far—such as Kenyan children from the Kalenjin tribe—develop into excellent runners. Kalenjin children, as a practical solution to their remote location, run an average of eight to

Figure 3.3 Young whooping cranes complete their first migration following an ultralight aircraft piloted by Operation Migration, www.operationmigra-tion.org. *Photo by Tim Ross, public domain via Wikimedia Commons.*

twelve kilometers every day, to and from school, to and from the river to bathe, and so on. They have a long-time tradition of long-distance running related to their cattle economy, as well as a culture that fosters what psychologists call a strong "achievement orientation." In a context where other careers are limited, and running already is common practice, and people grow up with a desire to do well and even better than others, the lucrative lure of the global athletic arena draws many children into training. Indeed, coaches can adopt a sacrificial attitude, pushing young runners to the limits of endurance and promoting the cream of the crop. It should therefore come as no surprise that Kalenjins and other Kenyans do so well in long- and even middle-distance running competitions (Shenk 2011, 102–105).

While some might like to write this off to genetics only, professional athletes themselves know better: a standing in-joke regarding how to hobble the Kenyans involves buying them school busses (Shenk 2010, 105). This would alter their developmental exposures, affecting their average phenotypes so that the Kalenjins would run more like members of non-running populations.

The epigenetic mechanisms underlying the manifestation of running prowess in Kalenjins per se have yet to be established. However, as David

Shenk has noted in his discussion of all this, based on extant science it seems clear that a genetic response, triggered epigenetically, leads to increases in certain structures within and around skeletal muscle cells that affect their stamina, given running's exertions. Likewise, muscle fibers grow in size and strength due to certain kinds of responsive "upregulation." Moreover, muscle fibers can change themselves from 'fast twitch' to 'slow twitch' as required (a higher proportion of slow fibers benefit marathon runners and cyclists; more fast twitch fibers benefit sprinters). While our plasticity does have pre-set limits, our muscles, says Shenk, "are designed to be rebuilt" in response to training (or inactivity): "At any given time, each muscle is adapted to a status quo of activity and exertion" (2011, 310, 312). Noting that one group of star athletes has more fast or slow twitch muscle fibers than another therefore means little; the distribution reflects their hard work and dedication to training rather than being a 'gift' that allows them more wins.

Even later exposures can make a phenotypic difference. Take London taxi cab drivers who, at least prior to computer-assisted navigation, had to memorize vast quantities of geographic knowledge to be licensed: there are 25,000 streets intersecting at various angles within six miles of Charing Cross Station alone, and many dead end. In learning London's lay-out, drivers changed the shapes of their brains. We know this because Eleanor Maguire compared experienced cab drivers' brains to those of a non-driver population. The cab drivers' brains had notably larger posterior hippocampi. Brain related variation existed not just between populations, but within the population under study, also: the longer a driver had been in the business, the larger his or her hippocampus was. The latter fact helped Maguire rule out the possibility that people with outsized hippocampi were simply attracted to the cab driving business to begin with (Shenk 2011, 34–35). Further research by Veronique Bohbot and others has shown that dependence on computer-assisted navigation is in fact associated with hippocampus shrinkage; populations that commonly rely on GPS or 'sat nav' may thereby be placing themselves at risk for relatively more cognitive problems later in life than those who still use spatial cues to get around (Raymond 2010).

Genes are switched on or off, silenced or promoted, by the environment via the epigenome. In other words, so-called instincts, skills, and certainly our preferences are shaped, this way or that, depending on environmental conditions and what they do to our epigenetic and, thus, our genetic systems. As Ridley notes, genes are "not just units of heredity… they are themselves exquisite mechanisms for translating experience into action" (2003, 275); they are "the servants of experience" (280).

Transgenerational Changes

The changes we have so far discussed in this chapter happen in a lifetime. They are developmental and not inheritable genetically—at least not in any direct sense. Having said that, maybe we haven't seen inheritance because we've been looking in the wrong place for it, or at the wrong scale.

Long before Darwin and others described the laws of natural selection, many people believed that changes brought on through experience in one generation could be passed to the next. For example, if a person who is not naturally strong works hard to build up a fit and muscular physique, his or her children, too, may be strong and fit. Or, for example, giraffes that stretch their necks by reaching upward have babies with longer necks than those born to giraffes that do not stretch. The idea that acquired characteristics can be inherited is called **Lamarckism**, after Jean-Baptiste Lamarck.

When systematically tested, the idea that acquired characteristics can be passed along falls short. For instance, a blacksmith who develops his or

Figure 3.4 *Village Blacksmith.* Rudolf Epp, steel engraving after an original painting, c. 1890.

her arm muscles through hard work at the forge may have children who end up with strong arms, too, if they take on blacksmithing (see Figure 3.4). However, the child of the blacksmith who goes off to work as a movie director will not have strong arms. The blacksmith's acquired strong arms cannot themselves be inherited. It is true, of course—as we saw for chest growth in higher altitudes—that any inbuilt propensity the blacksmith may have had for greater arm muscle development than normal could have been inherited—but his or her children's muscle development itself is not a foregone conclusion.

The Lamarckian theory of inheritance fell flat for more than just the lack of available evidence to support it. It had a competitor: the theory of natural selection. The latter easily and cleanly explained how characteristics were passed along without anyone having to acquire or work at developing them. Take, for example, the evolution of those long necks in giraffes. According to natural selection, those born with slightly longer necks than others are at a natural advantage when leaves are high up. Being better fitted to an environment where leaves are high, they simply have more offspring than those that, with shorter necks, cannot reach to eat as well. The next generation, therefore, is, on average, slightly taller—but intergroup variation still exists. Active self-development is not a part of the selective process, because it works on pre-existing variation within a population.

The Darwinian vision displaced Lamarckian and other ways of thinking about human evolution in a way that was as revolutionary in its time as the epigenetic advances now being seen. Epigeneticists may in fact help rehabilitate Lamarck's reputation. It turns out that some changes to our genetic material acquired in the process of epigenetic events may actually be passed along to offspring.

The science here is in its infancy, but scientists have documented, for instance in the work of Marcus Pembry and colleagues, how preteen cigarette smoking in one generation of fathers affected their sons. Specifically, smokers' male progeny were fatter than the sons whose dads did not smoke in their pre-teen years. Girl offspring were not affected. The group also examined records from 1890, 1905, and 1920 for an isolated community living in Northern Sweden. For this analysis, they explored not smoking but food supply in mid-childhood (that is, prepubertally), which it turned out was linked, for men, with grandsons' mortality risk; for women, with granddaughters'. It is the case that the researchers did not know how much food each father actually took in; they only had historical records regarding supply, which was in fluctuation in 1890 and 1905 and not in 1920. The effect was not seen in 1920, however. That and the fact that the associations in the 1890 and 1905 cohorts were sex-linked, as well as the similar finding for the sons of the smoking pre-teens, suggests that the proxy measure provides a valid way to demonstrate that some environment-linked pre-pubertal epigenetic changes can be passed along (Pembrey, Bygren et al. 2006; see also Whitelaw 2006).

The shift in thinking that conceiving of epigenetic inheritance demands may lead to a new way of talking about human evolution. For a start, in light of our expanding understanding of the epigenetic processes and genetic regulation's documented role for a small array of developmental adjustments, some thinkers have started to call this process 'epigenetic adjustment' or 'epigenetic change' instead. In addition, some have added another category entirely to describe intergenerational transmission of acquired changes in our genetic material (see Table 3.2).

It would be convenient if the latter category could be more simply labeled as 'epigenetic adaptation'; only the changes that were adaptive in the original generation may be maladaptive in the next. For instance, it seems that being able to survive on less energy in a time of famine (when

Table 3.2 Genetic Adaptation, Developmental or Epigenetic Adjustment, and Transgenerational Epigenetic Change

	Genetic Adaptation	Developmental or Epigenetic Adjustment	Transgenerational Epigenetic Change
Cause / Mechanism	Natural selection: Lower mortality and higher fertility rates among subpopulation with advantageous trait	Prenatal and childhood exposure: Physical plasticity responsively optimized during organism's development	Epigenetically induced changes in genetic structure are duplicated in genetic material passed on via reproduction
When realized	Next generation	This generation	This generation and the next
Directly heritable	Yes	No	No

food supply is low) may lead to epigenetic changes affecting one's ability to store fat that are maladaptive in the next generation if food becomes more easily available again or even overabundant (see Chapter 12). The epigenetics chapter that all next generation human evolutionary biology texts need to include has yet to be definitively written.

Paradigm Shifts

A **paradigm** is a theoretical framework that guides scientists in deciding what questions to ask, and forms a lens through which scientists interpret data. Questions regarding Lamarck's repudiation notwithstanding, the exciting discoveries in epigenetics heralded a radical change in scientific thinking—a **paradigm shift** in scientific assumptions regarding the relationships between species as well as concerning developmental change. Further, the debate over whether our genes or our environment contribute more to human biocultural diversity has shifted to emphasize discussion of the mechanisms by which genome, epigenome, and environment interact.

This is not the first time science has experienced such a transition. Some classic examples of changes in scientific viewpoint include when science came to accept that the earth is not flat, the sun does not revolve around the earth, and the earth is much, much older than we originally thought it to be. The revolution begun by the idea that the origin of speciation lies in a process called natural selection and that humans and chimpanzees as well as mollusks and fruit flies and even peach trees all share a common ancestor provides another good example of a jarring transition. All of these ideas—so commonplace now and so widely accepted that they have become axiomatic—were incredibly revolutionary in their times. Although data to support them were there, these ideas could not be accepted until a critical tipping point was reached.

Thomas Kuhn, a physicist and historian of science, described the radical nature of such transitions in *The Structure of Scientific Revolutions* (1996[1962]). Prior to every revolutionary transition—every paradigm shift—scientific practice is what Kuhn called "normal." What he meant was that it is unremarkable, driven by a shared, paradigmatic theory about the world. During **normal science**, most scientists in the community dismiss any contrary results or findings as due to faulty measurement or some kind of mistake in the analysis. A piece of dust must have been in the equipment, or the calculations indicating that, say, the earth is round, must have been in error. Investigator bias, too, may be suggested. Thus, when some members of one group exhibit a trait thought only to belong to members of another, and when not all members of that group can even be shown to have the defining trait, nay-sayers still clinging to the idea of set-in-stone human divisions may seek to rationalize this as merely reflecting the researcher's pre-existing viewpoint or shoddy investigatory skills.

Aberrant results are explained away or even ignored so as not to interfere with the daily work of science, and so it goes, until there are just too many anomalies or contrary results to rationalize away. Then, crisis ensues. Core beliefs are shaken as assumptions about the world reveal themselves as wrong.

When well-enough challenged, then, paradigmatic or standard views and normal scientific practices do change. Such change occurs radically and abruptly rather than gently and gradually. In other words, science does not progress at a gradual pace but moves forward in bursts as the growing unrest regarding contradictory data spurs what Kuhn terms a "scientific revolution." The old paradigm then is relatively quickly overthrown and a new paradigm takes its place. Soon, things settle down and another phase of normal science ensues. Disciplinary purviews are redefined; sometimes whole new disciplines or sub-disciplines emerge. Researchers carry on, now seeing the world through the new theoretical paradigm or lens. This is what recently happened when epigenetics arose. We now take human biocultural diversity to be a much more fluid and contingent product than scientists previously thought.

This is an advance, but it still bears noting that theoretical paradigms bias research, no matter how much better than previous paradigms they may be. In his book, Kuhn referred to the widely-known duck-rabbit image to convey this. In reference to the image (see Figure 3.5), Kuhn argues that, in a nutshell, if one has been told (or takes up the theory) that the image is a duck, then a duck is what one will see. Those expecting to see a duck ask duck-type questions; they may seek to measure bill size, or buoyancy, or feather softness. Conversely, those who approach the image with the theory that it is a rabbit ask questions relating to fur thickness, ear length, or maybe breeding habits. Duck-theory people fail to attend to the image's rabbitness, and rabbit-theory people do not see its duckness; all will fail to ask questions that fall outside of the box for them. Therefore, although people holding either theory in exclusion may do some excellent research into either ducks or rabbits, as the case may be, they will never be able to truly and fully accommodate the data. By highlighting the duck-rabbit figure, Kuhn helped readers to have the experience of shifting paradigms and to appreciate the degree to which paradigms, for better and worse, set our agendas.

This is not to say that Kuhn dismissed the notion of progress through science. Clearly, science does advance and discoveries are additive and

Figure 3.5 Duck-rabbit. J. Jastrow (1899). "The mind's eye." *Popular Science Monthly,* 54: 299–312.

even multiplicative. Who reading this book would not prefer today's treatments for, say, smallpox or cholera over the treatments available in the 1800s? Kuhn is not describing the simple replacement of one set of biases for another. Rather, he is pointing to the conservative disposition of science in general. The willing and sometimes automatic adherence of scientists to shared paradigms can muddy our view of the laws of nature, including those relating to heredity and evolution, and to human variation. Science is, after all, a cultural product.

What Makes a Square a Square?

We began this chapter by asking how genotypes are translated into phenotypes. Sure, the proteins that genes code for are built and assembled into body parts or process-supporting biochemicals. However, like baking a soufflé, turning a recipe into a body is a complicated and contingent act. As we have now seen, the phenotype is in fact the result of the concerted interplay of genotype with epigenome and environment. Put into an equation: phenotype = (genotype + epigenome) x environment.

Another relevant puzzle—and one that is implicated in the entire content of this book—is whether nature or nurture contributes more to who we are, or to the ways in which populations bioculturally differ. However, that question leads to a dead end. Just as it is illogical to ask which contributes more to a square's area, the width or the height, so, too, it is nonsensical to pit nature against nurture or vice versa and search for a victor. Like the square's height and width, nature and nurture work together to produce the wondrous diversity we see in **humanity**. Nature (biology, genetics) *and* nurture (culture, the environment in which we are conceived and raised) both work together to shape us, and we can see the result in both individual- and population-level biocultural diversity.

Dynamics of Human Biocultural Diversity: A Unified Approach by Elisa J. Sobo, 51–67.

4

Emergence of Culture and People Like Us

This chapter prepares you to:

- Define culture with reference to its key features and adaptive function
- Define and discuss cultural relativism (in contrast to ethnocentrism), using examples
- Describe culture's evolutionary emergence with reference to the synergistic interaction of our evolving capacities for motor skills, empathy and cooperation, and symbolic communication
- Describe key archaeological indicators of the Cultural or Paleolithic Revolution

I have been talking about adaptation as a bio-genetic process. In this chapter, I want to discuss another option humans have in adapting to environmental change: cultural adaptation. Culture helps us to adapt to various geographically-distinct environments by providing us with a means to invent and then pass on customs pertaining to survival-related needs for food, clothing, and shelter. Moreover, culture helps communities to stick together or cohere. It gives life meaning; but I will get to that part later, after first defining culture and then reviewing how humans got to be culture-bearing beings.

Culture's Adaptive Function

Geography had great influence on culture prior to recent times (and remember that 'recent' is a relative term, us humans having been around in modern and culturally-able form for maybe 100,000 years now). That is, our initial food-getting or subsistence strategies as well as the materials used in the clothes we prepared and wore (if any) and the shelter we took (if needed) were geographically determined originally. We had only the resources of our habitats to exploit. The environmental conditions they entailed determined our survival needs and therefore our immediate cultural aims. Where it rained, shelters would need waterproofing. Where fish lived, we might develop techniques for fishing. Where arctic winds blew, we would need warm clothes, and so on. The ability to adapt using culture helped support our spread from Africa, where the human species was likely born and our capacity for culture likely evolved, to populate most of the earth's land mass.

Culture's Features

Commonly, when we characterize culture in introductory anthropology courses, Edward B. Tylor's 1871 definition is invoked: "Culture … is that complex whole which includes knowledge, belief, art, morals, law, custom, and any other capabilities and habits acquired by man as a member of society" (for example, as cited by Langness 2005, 25). This definition is offered up in part because it is thought to be the first English-language definition of culture; further, most theorists agree that it has basically withstood the test of time and is as good as any other definition around. It works because it highlights three key features of culture, particularly in the line mentioning "habits acquired … as a member of society": culture is (1) learned or acquired, (2) shared socially, and (3) sticky or habitual. Indeed, learned, shared culture becomes so deeply internalized as habit that we often forget it is there, and in fact we may never 'see' it until we knock up against someone with another culture—someone with learned, shared habits that differ from our group's.

This brings up a fourth point implicit in a part of Tylor's definition not shown above: culture is (4) self-centered. Our deep internalization of our culture leads to 'ethnocentrism,' defined in Chapter 1 as using our own cultural standards to judge other peoples' practices.

Tylor in fact referred to "civilization" in part of the passage that I have not included, and positioned his own culture at the top of a progressive evolutionary ladder. This was quite common in Tylor's day, as cultures were thought to all pass through a particular progression of evolutionary stages (for example, from savage to barbarian) until they reached the pinnacle of enlightened development, normally and ethnocentrically benchmarked by the anthropologist's culture.

However, things change. In the United States, anthropologists began to see that different societies had different histories and their cultures were therefore unique and particular. The technical term for this viewpoint is **historical particularism** and it is most associated with Franz Boas, one of America's anthropological pioneers and, not coincidentally, an immigrant marked by his own ethnic differences from the mainstream U.S. Anglo elite.

Boas was appalled by the misuse of science to support racism, which was essentially what most of the evolutionary ladders ranking other peoples' cultures were promoting by the end of the nineteenth century. Boas counterattacked, arguing (and rightly so) that environment influenced bodily form, that there was no such thing as a 'pure' race, and that 'racial mixing' was not harmful. His cultural research directed attention away from **unilineal** (one-way) evolutionary thinking that placed the ruling culture at the top toward a paradigm that allowed for many lines of cultural evolution. Overall, Boas's work highlighted the need for **relativism**, which evaluates the ideas and practices enacted by members of a given culture by that culture's own standards. They are understood neither as they rank on a spurious or baseless developmental scale nor from the ethnocentric viewpoint of the foreigner-theorist, but rather in their own context.

The paradigm shift in anthropology promoted by Boas's work is clearly seen in the changes museum curators made in response to his insights. Prior to Boas, artifacts were generally arranged, no matter where from, by category. Like objects were grouped together in a bogus unlinear progressive sequence running from 'savage' to 'civilized.' Such an arrangement showcases different approaches that humans might have to the same type of adaptive challenge (meat procurement, say, which might take place today using bows and arrows, or high-velocity computer-assisted rifles). Yet organizing artifacts out of cultural context detracts from museum-goers' (let alone scholars') capacity to develop a holistic understanding of a given artifact's place in the cultural world from which it was wrested.

An Inuit basket, for instance, could be displayed with other baskets, say, baskets from around the world. It could be placed in between those that are more and less complexly or intensively crafted, for instance. However, displayed in the context of other Inuit artifacts instead, such as those used likewise in food-gathering and preparation, the Inuit basket—and the artifacts now around it—become culturally meaningful. Museum goers now know more about the basket's purpose; they can better imagine what it might be like to use it. When shown on its own or with other baskets only, all that remains uncertain.

Context-sensitive relativism was what Boas demanded. He even modeled himself how certain artifacts were to be used or worn by display dummies so that they might be seen not only in context but also as if in

Figure 4.1 Franz Boas, modeling for an exhibit as "Hamats'a coming out of secret room" in a Kwakiutl Indian ceremony. *Photo courtesy of National Anthropological Archives, Smithsonian Institution, Negative MNH 8304.*

use or action (see Figure 4.1). Such museum display tactics allow the viewer better entry, however virtual, into other cultural worlds.

The Emergence of Our Species

We have talked about the why and what of culture, and the importance of context. We still need to talk about when and how culture came into being. It is an evolved capacity—one that only became possible after a number of genetic changes had built up, leaving humans better able to make things and to communicate using language and, more importantly, to cooperate in reaching shared goals and to identify with the group.

Before we talk about culture's emergence, however, it will be helpful to quickly review the emergence of the broader group that humans fit into. We are firstly part of the genus *Homo*—literally, 'man' (a **genus** is a set of species sharing particular biological features). *Homo* first appears in the fossil record about 2.5 million years ago, as the species ***Homo habilis*** (species generally are identified in scientific writing with the genus named first, due to the important information that the genus name conveys). *Homo habilis* means handy man; this species was the first ever known to systematically make and use stone tools. These tools were quite basic—rock hand-axes for smashing or crushing or scraping—but they were tools nonetheless. Their appearance in the archaeological record marks the entrance of the genus *Homo* and the beginning of the early or Lower Paleolithic (literally, 'old stone') era.

Homo habilis certainly made less invasive tool-related modifications to its environment than we do today. Nevertheless, use of tools intensified the species' niche construction (niche shaping) activities in a trend that would slowly gain speed later. You see, something else interesting marked this new species and the genus it represented as different: *Homo habilis* also had a larger brain—relatively speaking—than any similar species at the time or prior. What might explain higher fertility and lower morbidity rates in bigger-brained population members than others? Better thinking: even a slight enhancement in brain power would come in handy in fending off predators, one of the major environmental pressures faced by *Homo*—and indeed all primates—at that time (Fuentes 2009).

Homo continued to evolve, but the next notable step-change in *Homo* fossils did not come until one million or so years later, when **Homo erectus/ergaster** entered the scene. This species really optimized its bipedalism. Earlier *Homo* species could walk; even non-human primates are known to stand on two feet sometimes—for instance, when gathering food. However, this new species walked all the time, habitually, and more effectively. Altered body proportions reflect this. *Homo erectus/ergaster* was taller and had longer legs and shorter arms than other *Homo* species. Due no doubt in part to its bipedal efficiency, this species is the first found in locations beyond Africa—a topic we return to in Chapter 5.

In addition to its more truly upright posture, *Homo erectus/ergaster* also had a more protrusive nose. This may have been selected for in support of the kind of breathing that more active bipedal locomotion, or running, requires. Running in turn may have provided an advantage in hunting, allowing individuals to chase after prey that might otherwise escape. Also, it could be an advantage in the race to avoid being chased down and eaten oneself (Fuentes 2009).

As the threat from predation declined, another selective pressure was on the rise: babies became a bigger burden. *Homo erectus/ergaster* had a still bigger brain—and so a bigger skull top or cranium—than *Homo habilis*. Babies now had to be born less mature or their heads would get stuck, killing them and their mothers. This meant that infants, once born, took somewhat longer to become independent enough to survive on their own.

The energy costs reproductively mature females bore in carrying and feeding their now-more-dependent offspring must have been offset by a turn, however slight, toward more food sharing and group food gathering efforts. This would have entailed more cooperation and cooperation on a different order than is seen in other primates, but it is important not to project back in time today's human qualities. Cooperation at this stage was still unlikely to entail the sense of mutuality that we experience when cooperating today (Tomasello 2011).

Before we come to that, however, it is worth noting that bigger brains and taller bodies ratcheted up the need for food overall. Diet evolved concurrently, and tools became somewhat more sophisticated in concert.

Not to belabor the point, but one adaptation did not follow the other in simple linear fashion; numerous connections within and between systems were implicated. The adaptations that contributed to *Homo's* evolution linked up with each other through multiple, multidirectional feedback loops. Locomotion, breathing systems, brain size, food sharing, diet, tool-making skills, and so on all changed in self-reinforcing—and sometimes self-limiting—ways.

For example, changes in the diet that supported advantageous changes in brain size may in turn also have supported intestinal shortening and tooth size reduction. These changes would have been largely due to the advent of cooking, which necessitated first learning how to capture or control fire (making fire came later). Cooking predigests food in effect, making many nutrients and certain energy sources more available to the body. Individuals with smaller teeth or shorter intestines would have died off in earlier days due to an inability to actually get the nutrition they needed from then-available foodstuffs. Now, however, in a niche constructed to include cooked food in the diet, small-toothed, short-intestined individuals were more likely to survive. Genes with the potential for expression as smaller teeth or shorter innards would therefore be passed along into the next generation's gene pool at a higher rate than before. Concurrently, people with larger teeth and longer guts would no longer have a selective advantage.

Interconnected changes like these, and others, accrued as time passed. Various new *Homo* species emerged, including, about 500,000 years ago, archaic forms of bigger-brained ***Homo sapiens***. The subspecies ***Homo sapiens sapiens*** (anatomically modern humans, the group we fit into) entered the world about 200,000 years ago. Scholars today hold a number of ideas about the links between diverse early *Homo sapiens* species and sub-species, and about the links from them to us. All agree, however, on some anatomical facts: *Homo sapiens sapiens* can be distinguished by our less-robust, highly domed skulls, and vertical foreheads. Our noses project more than those of other *Homo sapiens,* and we have bony chins. Also, we have smaller teeth and jaws, and our skeletons are lighter overall.

Sapiens is Latin for wise or knowing. From its emergence, this new, bigger-brained species of human was smart in a way that others in the genus *Homo* hadn't been. *Homo sapiens sapiens* was even smarter still: the brains of this species were—are—different from other *Homo sapiens*, not so much in terms of size, but in terms of what they could and can do. *Homo sapiens sapiens* were primed, through a rewiring or reorganization of the brain rather than simple growth in its size, to learn—to carry and enact culture. They were ready to imbue the world with meaning and to talk about it; they were ready to experience, subjectively, a state of being 'we.' When this happened, we went from being simply anatomically modern to being what has been termed 'behaviorally modern.'

Setting the Stage for Culture through Tools, Talk, and Teamwork

As *Homo sapiens* begin to spring up in the fossil record, the archaeological record also changes. By this point in our evolutionary history, it seems that some groups may have cared for the aged and infirm, and even buried their dead (or some of them). That they did so is significant. As noted earlier, culture provides humans with a way to adapt to various geographic habitats without waiting for genetic adaptation to kick in. Beyond brute survival, however, culture provides the glue that holds human communities together. It makes meaningful social life and real teamwork possible. In the titular words of Walter Goldschmidt, culture's emergence provided our "bridge to humanity" (Goldschmidt 2006). Person-to-person or intergroup connections made via language and other ways helped our evolving ancestors cross this bridge.

From Tools to Talk

As Goldschmidt notes, our capacity for talking to one another and our capacity for making and using tools seem to have evolved hand in hand. Both involve classifying things, connecting them (concretely or mentally), and dexterity. To make a tool, you must have in mind ideas about what types of materials might be best, and how to modify them to gain the most effective tool action. More complicated tools may require you to prepare materials in some way before tool construction. For a multi-part tool, you must consider how parts can be connected. Making and using tools can require fine motor coordination and dexterous fingers and hands.

Recent discoveries regarding the brain suggest that the manual dexterity necessary for making and using tools may be intimately linked not only with the development of gesture language but also (although it took longer, evolutionarily speaking) with the development of the parts of the brain implicated in the manipulations of lips, teeth, breath, and tongue that are necessary for speech. The connectionist actions entailed in tool-making, gesturing, and talking, and the cognitive mechanisms behind them, are in some ways very parallel. In other words, we use very similar skills for making tools and communicating, and it is likely that they all evolved in tandem.

Humans can inform other humans of things that they think might be helpful, using speech and even gestures (the emergence of which, hand in hand with tool making, no doubt scaffolded the development of spoken human language). In other words, humans can share information for the other person's benefit. Such information sharing would not have been seen in our early *Homo* ancestors; it still is not really seen in non-human primates today.

Great apes, for instance, have trouble understanding that when a human points a finger at something the act is meant to convey information to them (such as "There! There is some food that has been hidden!"). They do not have the ability to identify with the pointing individual and, moreover, to trust her or him (Tomasello 2011). This may seem surprising in light of the fact that primates live in groups. Remember, however, that among both our ancestral species and non-human primates today, group life was and remains pragmatic. Member behaviors were not cooperative in today's sense; they were individualistic and even exploitative (Tomasello 2011, 5). Early *Homo* groups did travel and eat together. Nevertheless, as remains mostly true for today's non-human primates, this was basically just to help each other avoid being taken down by predators: there is safety in numbers.

Humans today are different. Our ability to identify with someone—a teacher, for instance—is keen. It has to be: our really big brains mean babies are born much more immaturely than ever before. While non-human primates (and other creatures) move from infantile dependency to the so-called 'juvenile' stage without a pause, humans have childhood. Childhood per se is unique to humans, whose brains have so much to accomplish in becoming mature. This includes more than just physiological growth: we can no longer maximize our environment without being told or shown how to do it. As part and parcel of having childhood, we are primed to teach and learn.

Inter-subjective Mental Connections

Sometimes, when we observe someone doing something that we would imitate or that we can see ourselves doing—say, picking up a particular shell and rubbing it just so, to create a hole through which to thread it—our own brains respond as if we already are engaged in practicing that action, too, although we only are watching. That is, some of the neurons or brain cells that control specific actions can fire simply because we are thinking about that action. Once discovered, these were aptly named 'mirror neurons.'

Interestingly, mirror neurons seem to be concentrated in a region of the brain adjacent to the region most actively associated with language production. This points to further synergy between use of the hands—here, for making communicative hand gestures rather than (or,

Figure 4.2 A newborn macaque imitates tongue protrusion. L. Gross (2006). "Evolution of Neonatal Imitation." *PLoS Biol 4*(9): e311. doi:10.1371/journal.pbio.0040311.

more likely, in addition to) tools—and the emergence of language proper (Arbib, Liebal et al. 2008).

Much more research is necessary, and we need to guard against accepting a facile 'just so' story in which mirror neurons explain all. Nevertheless, neurons with this mirroring capacity seem to be part of one key mechanism for non-genetic information passage (Goldschmidt 2006, 31; see also Ridley 2003, 209–222). Their existence indicates intense connectivity in the brain.

Certainly, non-human primates have mirror neurons: these were in fact first described through brain studies done on monkeys watching other monkeys, and some humans, bring peanuts to their mouths in Giacomo Rizzolatti's laboratory (Blakeslee 2006). That non-human primates can mirror or imitate others overtly, too, is well-known (see Figure 4.2). However, human mirroring entails qualitatively and measurably greater sophistication—the kind it takes to enable and support culture.

The emergence in *Homo* of the capacity for the kind of learning that culture depends on may be related, at least in part, to a boost in the quantity and quality of mirror neurons thousands of years ago—itself of course the result of feedback coming from a variety of directions related to how our ancestors then lived. More, higher-quality mirror neurons might not only be helpful in supporting our quick uptake of motor skills, such as when we learn how to make or throw a spear, or soothe a baby, or weave a basket by watching another member of our group do so. Mirror neurons may respond, too, when we watch a film or play that moves us, or go to a favorite team's sporting event. They can fire when we see someone struggle with and nearly drop an armload of packages, or trip while walking across campus. They seem to be implicated, therefore, not only in

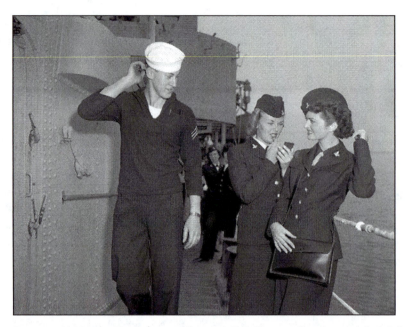

Figure 4.3 Marion Koopman, *right,* and U.S. Navy sailor Samuel E. Adolt mirror each other; they and Margaret Williams, *center,* are on board the Fletcher-class destroyer USS Uhlmann (DD-687) at Terminal Island, California. *Photograph: Los Angeles Times, 1950; UCLA digital collections photo ID: uclalat_1429_b104_66045-2).*

motor learning but also in physical acts of empathy—of placing our own feet in someone else's shoes or of feeling a connection (see Figure 4.3).

In this way, mirror neurons also may have supported efforts to transfer emotions, or feelings, from person to person. This kind of empathic transfer can help to create and maintain community bonding. We must not underestimate the importance of such bonding: without it we cannot cooperate.

Today humans know that 'two heads are better than one' and 'many hands make work light.' Rather than evolving two brains, however, or two sets of hands, behaviorally modern humans evolved the ability to cooperate very, very well. For example, when gathering food, we may divvy up and even rotate through roles (trail cutter, child carrier, resource locator, food extractor, and so on). On a hunt, some people may climb trees to flush out prey while others act as shooters. Before heading out for plants or animals, groups agree on joint goals. Their work is coordinated and synchronized.

Now is a good time to mention that human eyes today have big, bright white areas (sclerae) surrounding the pupil and iris; non-human primate eyes do not. Eyes with big, bright whites—human eyes—are much easier to follow than the eyes of any other organism. We do not know when our ancestors got their big whites, but we do know that, in light of how natural selection works, highly visible whites would have only been selected for when they did not disadvantage the bearer. For example, an animal's big whites could allow others to spot food it is eyeing and to pick that food and eat it before the animal has a chance. At a high enough frequency, such exploitation could lead the organism with the visible sclerae to starve and die off prior to reproducing—prior to passing the genetic basis for the bright whites along to the next generation. Selfish exploitation is the kind of thing we might have seen in early *Homo*. In *Homo sapiens*, however, knowledge of where another is looking must have been used also, and more so, to coordinate interaction in a way that increased the fertility of all whiter eyed parties (*Economist* 2006).

Successful cooperation depends upon empathy, mutuality, and, more specifically, as Tomasello notes, "shared intentionality." It takes a certain kind of intelligence to be able to relate to the concept 'we' or to enact 'we-ness,' as people do when dancing a tango or helping to bring in the harvest. One must be able to understand the actions of others as tied to certain intents or goals, form with others shared objectives or conventions, locate and identify what others are attending to, direct the attention of another somewhere else, and so on (Tomasello 2011).

Non-human primates (and by extension, our ancient ancestors) do not do these things. With some modest exceptions their work only counts as group work because individuals share proximity or are spatially located near each other as they forage, eat, and so on. Among humans, however, a life of minimal cooperation was not sustainable. Human groups that cooperated better simply had a better chance of survival in light of the new landscape of selective pressures that their evolved traits now exposed them to. Those who could really cooperate had a selective advantage. Without cooperation, culture not only could not work but it would never have had the opportunity to arise in the first place.

The Role of Cooperative Childcare

One push for cooperation was infant dependency, which had been on the rise ever since our brains began expanding our skulls. Earlier, I mentioned that food sharing and group food gathering would have had to increase. Here, I will add that other changes, related to bipedalism, caused infants to be born basically rear-facing instead of facing up or forward, as non-human primate infants are today. This change, says Wenda Trevathan (1999), made unassisted childbirth a bit harder, for instance making babies' necks more vulnerable to damage as mothers reached down to guide them up and out. It was not—and is not—impossible, but having someone there to help 'just in case' doesn't hurt. Accordingly, while non-human primates seek solitude upon first feeling labor contractions, humans seek the company of others. This propensity was selected for: those females who first sought assistance as a response to labor contractions had more offspring survive to themselves reproduce.

At first, the urge for companionship during labor may have just been a variation in how individual organisms responded to pain. Later, our ancestors would have evolved the capacity to realize, intellectually, what challenges might be entailed in the birthing process. They would have evolved the capacity to pass such information on to others as well as the capacity to absorb (learn) such information, so that personal experience was not the only teacher. Once natural selection had led to this ability to communicate, they would have been motivated by the anxiety that communicated (shared) knowledge generated to ask for help, and to provide it (Trevathan 1999). Selected for, the predisposition toward social support during birth surely had a hand in driving then-emerging cooperative—and information acquisition—skills even higher.

Increasing rates of birth attendance also would have been coupled with, and in turn helped ratchet up, increasing rates of the behavioral strategy called **allocare**. This entails providing childcare services to children who are not one's own biological offspring ('allo' means 'other' so 'allocare' means, literally, 'other care,' as opposed, for example, to being cared for by your mother). Allocare can entail anything from occasional babysitting to the longer-term arrangement between a nanny and his or her charge, or a foster family and foster children. The extended family often provides allocare. Another example of allocare not infrequently seen in the ethnographic record entails breastfeeding a child that did not come out of one's own womb. Our species seems to have practiced cooperative child care of this kind since (and likely since before) its very inception.

Allocare provides the biggest selective advantage in communities where food might be scarce. It happens elsewhere, too, but, when getting enough food to feed the family is a hard thing to do, having other people help care for the children so that able-bodied adults can go out and collect more food is all the more advantageous. Individuals who may not be able to walk as far (for example, grandparents) can make an important contribution to group survival by pitching in with allocare. Moreover, older siblings who help take care of children learn what we would term 'parenting skills' in this fashion (see Figure 4.4).

Allocare seems to have mutually reinforced the natural selection of other capacities that support productive cooperative interaction. Think of the advantage gained by infants and children who are better at getting people to meet their needs. By making certain faces or gestures, young ones provoke allo-maternal or allo-paternal responses in people charged with taking care of them. Children who are better at figuring out when, where, and how to make such faces or gestures and who therefore are better at getting others to take care of them—and to teach them things— simply have a better chance for survival. Likewise, children have a better chance for survival if they are able to discern what potential caregivers are thinking or feeling: those children who are better skilled at reading the emotional states of adults and older siblings around them can target requests for food, hugs, care for an injury, or whatever it is that they need to those more likely to be responsive.

Like birth attendance, then, the early practice of allocare was a likely contributor to the natural selection of perceptive and expressive capacities necessary for productive group life. Like tool making and talking (and all that these entail), it had knock-on humanizing effects, reinforcing the emerging brain structures responsible for enabling and promoting human connectivity. However, something more had to enter

Figure 4.4 An older child providing allocare. Soriya plays Bong Srey ("big sister" in Khmer) to her younger sibling while her parents are out working. *Photo courtesy of Riverkids Project.*

our behavioral tool-kit before the real transition to humanity—people with full-blown culture—could happen.

Ritual and the Cultural Revolution

Given the evolved capacity of humans to connect, verbal language is an excellent tool for transmitting the learned, shared information that comprises much of human culture. For instance, we can tell each other how to make regionally-relevant houses or clothing. Moreover, we can gossip, which conveys essential information about norms and values—indeed, 70 percent of all talk is actually social gossip (Dunbar 1996).

However, as Goldschmidt explains, there are some "problems with the way language handles the complexity of the world" (2006, 33). Yes, language is excellent for conveying technical information, such as how to make a particular type of blade from a particular type of stone, or that you need lunch, or regarding how to—or not to—behave at the company picnic or in class or at an uncle's wedding. In and of itself, however, verbal language is not always so good for conveying or transferring sentiments or feelings from one person to another. For this, day-to-day verbal language needs embellishing, such as through lowering your voice or clasping the hand of the person you are speaking to. Better than these techniques, there is the language of **ritual**—of action sets performed for their culturally relevant symbolic value.

Ritual communication includes the more kinesthetic or bodily communicative mediums of song, dance, and prayer as well as systematized

Figure 4.5   Ritual event: "The Thanksgiving Dinner," Corner of Grand & Allen Streets in Ridley, New York, 1870. Artist unknown, *public domain.*

gestures or actions such as eating together, and so forth. Ritual allows us "to create the emotional cohesion that makes for effective and continued collaboration and to reduce tensions between potentially hostile persons and groups" (Goldschmidt 2006, 40). As Goldschmidt further explains,

> [Rituals are] multimedia events that make everybody involved feel the same way about whatever the group needs to feel about. If you want to psych up youths to go to battle, you get out the drums and gesticulate in unison.... If you want to solidify a group you eat together and sing and dance in unison.... Ritual communication stands in contrast to the engineer's logic of speech, supplementing it with what I call an 'aesthetic appreciation' that transmits feelings from one person to another—not information about feeling, but the feeling itself.... [Rituals are the] cultural invention that made culture work.... Tools give the cultural mode of life its survival advantage, and language is the device that made culture possible, but ritual was needed to hold the social fabric together. (40–41)

Rather than simply instructing, ritual motivates. It entices our social and cultural commitment. Ritual coerces through its ability to mobilize emotions and transfer them between members of a group (see Figure 4.5).

With ritual communication, it is not the explicit messages—for example, "Turtles are our clan's totem" or "This is the American flag"— that are of the most import, but rather the viscerally-felt confirmation that "Our clan (or society) is a wonderful group to which we should be loyal." Feelings are paramount. Paradoxically, as Goldschmidt notes, imagination and sentiment are essential social catalysts for us humans, although we often focus instead on our rational intelligence as the thing that makes us special (42).

The Full Bloom of Culture

Although we may have been capable of language with the emergence of *Homo sapiens*, we first see tangible evidence that humans are really attending to social connections about 75,000 years ago, give or take. This was about when full-fledged culture—the kind supported by ritual—seems to have emerged (traditionally, a date of 40,000 years ago might have been given, but evidence for an earlier emergence has been piling up at an extraordinary rate, and at this time 75,000 years ago seems the best approximation). In what, evolutionarily speaking, was 'all of a sudden,' we became truly human: we began to express ourselves symbolically, as through art and personal adornments (see Figures 4.6 and 4.7).

Figure 4.6   Front and back views of a female figurine in ivory from Kostenki, in Russia. About 25,000 years old. *Courtesy Olga Soffer.*

Figure 4.7   Shell beads from the 75,000 year old levels at Blombos Cave, South Africa. *Photo by Chris Henshilwood and Francesco d'Errico via Wikimedia Commons, CC-BY-2.5.*

All of a sudden, the archaeological record includes things like small statuettes, often depicting female figures. Less portable but no less impressive were the polychromatic cave paintings, for example of horses and bison, that also began to appear. In addition, we see string, thread, and sewn clothing as well as shell and tooth beads for adornment. More complicated tools arose, as did artifacts apparently meant to "support, defend, and amuse the deceased" (Ehrlich 2006, 170) in burials. Most theorists believe that such artifacts are sure signs that symbolic life was in high bloom.

This change in the archaeological record marks for many scholars our entry into the time phase known as the Late Stone Age or Upper Paleolithic. It reflects a blossoming of human self-awareness, and although by this time various *Homo* species had set out from Africa and begun to populate parts of Europe and Asia, this blossoming seems to have happened first in Africa. That is where the oldest cultural artifacts have so far been found (see again Figure 4.7). Therefore, with an exception soon described in Chapter 5, most scholars today agree that Africa was where behavioral modernity (that is, full-fledged cultural life) was born. Although various *Homo sapiens* groups may have had some rudimentary cultural life, the real thing—culture full-fledged—seems to have been enjoyed only by our particular subspecies: *Homo sapiens sapiens*.

The change in the archaeological record described indicates, too, that human beings had by now developed the capacity for spirituality and religious beliefs. Some say that without cultural belief systems to justify or at least make sense of the tragedies and trials that inevitably accompany human life cycles, the anxiety that must have accompanied our ancestors' increasing intelligence and self-awareness would have been too much to bear. In addition to providing a mechanism for achieving social cohesion, then, the meanings conveyed through ritual also provided a means to mitigate existential angst.

The transition to culture appears to have been made so relatively quickly and dramatically that many scholars speak of it as the **Cultural Revolution** or (to avoid confusion with historical events in China) the **Upper Paleolithic Revolution**. Explaining why the transition to culture or true behavioral modernity happened so fast is fairly easy: a threshold was surpassed. However, even if we agree with Goldschmidt (2006) that the emergence of ritual practice pushed us over the tipping point to engage in full-fledged cultural life, we still must ask why such a long gap existed between the evolution of our capacity for language as *Homo sapiens* and our engagement in a fully cultural life. Goldschmidt points to the lag time common between any technological advance and a cultural solution to the problems or changes that it brings about.

Think about the smart phone, and how we are still adjusting to the social changes it has wrought (the ability to send and receive messages instantly and globally, for instance, has affected both politics, as in the Arab Spring, and demographics, as in death rates related to texting while driving). Early innovations—including communicative innovations such as passing on information verbally—could only be optimized if we could maintain social cohesion, and it simply took us awhile to figure out how to do that. Once we could really leverage our existence in cooperative social groups, which we did through the communicative practice of ritual, we crossed the "bridge to humanity" described by Goldschmidt, and culture took off.

5

Global Expansion, Human Variation, and the Invention of Race

This chapter prepares you to:

- Outline the immediate evolutionary history of *Homo sapiens sapiens* and explain how the species populated the earth

- Recount the history of the race concept as applied to human subgroups and explain the difference between biological race, racialism, and racism

- Define geographic cline and explain the value of making geographic ancestry-based distinctions between subpopulations in certain limited circumstances

- Explain the genetic argument against race, with reference to genetic variation within and between diverse groups

The balance of evidence today indicates that the capacity for culture discussed in Chapter 4 emerged when our ancestors all lived in Africa. Since that time, we've spread out around the globe. The creep of humanity across the earth underwrote the emergence of a good deal of diversity in how geographically distinct groups look—as well as in things unseen inside of us. As interesting and important as these internal differences may be, until recently all that we humans could recognize were the superficial ones—the ones that affected the texture of our hair and the color of our skin, for instance. In fact, when most people think about human diversity, they still think about these kinds of superficial physical differences, broadly sorted into a handful of racial categories. That's too bad, because differences in eye shape, nose form, skin shade, and so forth represent only the tip of the diversity iceberg. Moreover, the way such features can be grouped together to separate us—their categorical utility—has been greatly exaggerated.

This chapter supports a more sophisticated understanding of human diversity by examining how the diversity of environments we met with when we populated the world affected various subgroups of our species differently—but not so much that separate races really were created. Recall that the bulk of tangible evidence suggests that by about 75,000 years ago, and likely even earlier, behaviorally modern humans—that is, humans with the capacity for culture—had emerged. This emergence happened in Eastern Africa. Archaic humans had lived in the region for much, much longer; in fact, African populations have the most varied gene pools in part because of the vast time span of their history.

Archaeological and human skeletal evidence also provides insight into the roads taken by some of these African human ancestors as they spread over the various continents on the globe. In the process of migrating, many of these small offshoot groups encountered environments quite different from those they enjoyed in East Africa. Some regions were colder, or hotter, or higher, or lower, or drier, or wetter, or varied on other dimensions from the environment in which the original human beings had done most of their previous evolving. Foodstuffs differed, too. To survive, some groups of humans adapted to new environments in ways that meant their bodies as well as their behaviors changed.

Importantly, as this chapter shows, most adaptive variations evolved independently. Skin color, for instance, need not—and often does not—correspond to height, eye shape, nose width or length, hair texture, blood type, red blood cell shape, or any other trait in a given population. Most traits associated with the so-called races do not clump together in predictable, stable patterns. Moreover, patterns that *do* exist correlate with geography, and generally with regions that are fairly limited in size, so that knowing the location of one's particular ancestral homeland, and its evolutionary pressures, has much more predictive value in regard to such traits than does one's so-called race. Biologically speaking, humans are of one race, not many.

The Spread of Humanity

It is by now well-established that *Homo* did most of its evolving in East Africa. However, the fine details of exactly when and how humans like us took over the world remain the subject of much debate. While results of these debates will not change the substantive content of this book, it still is worth describing some of the issues being debated, not least for the practice it gives us in weighing theoretical options.

Spreading Out from a Single Origin

The most widely accepted theory of how we spread out to occupy nearly the entire globe is known as the 'single origin' or, more commonly, the **out of Africa model**. We know that we walked out of Africa; scholars generally agree that we walked first to the south and west in Africa and then to the north, moving out of Africa through Southwest Asia as early as 90,000 years ago. By about 60,000 years ago, some humans had moved globally eastward, through Southern Asia, and then into Australia-Oceania and East Asia. By about 40,000 years ago, humans as we know them had entered Europe. Eventually, maybe 20–15,000 years ago, behaviorally and anatomically modern humans reached the Americas as well, having moved through Central Asia and then having crossed to the new continent at Alaska from what is now Siberia (see Figure 5.1).

Figure 5.1 Migration routes ancestral humans may have taken in populating the continents. 1=Homo sapiens; 2=Neanderthals; 3=Earlier species. *Adapted from Spreading Homo Sapiens by Magasjukur2 via Wikimedia Commons, CC-BY-2.5.*

Location-Linked Diversity: A Small Variation on the Basic Human Theme

Each of the new environments that we entered posed slightly different adaptive challenges, so new variations were selected for persistence in the human gene pool over the course of our spread. Our basic human nature—our anatomy and physiology, but also, and crucially, our brain-based intelligence and creativity, aptitude for linguistic and other symbolic expression, and capacity to empathize and form social bonds—was born in Africa. Through migration, however, we augmented our basic humanity with new and relatively superficial, region-based biological diversities. We adapted culturally, too. Most diversity is of course biocultural rather than simply biological or simply cultural. Here, however, we will focus on geographically based explanations for location-linked bodily adaptations.

For example, by the time humans had spread across the earth instead of all living in one limited spot on the globe, rather than everyone having darker skin, as was the norm in Africa, some populations had evolved skin that was lighter in color. Likewise, rather than everyone making lactase only in their first few years, those whose new environmental adaptations included reliance on milk-based foodstuffs not available in our earliest days in Africa evolved the ability to make lactase well into adulthood.

Importantly, non-Africans have only been non-African for a short amount of time relative to the time-depth of the African population from which they branched off, and those who branched off represented only a small part of the total African gene pool. Therefore, the degree of genetic variation in non-African populations is much lower than the degree of variation found on the African continent. Moreover, most genetic variation found outside of Africa represents a subset of variation also found within it.

Migration Waves

Migration did not occur, it turns out, in a single move. Many theorists believe, based on skeletal and genetic data, that small groups of humans left African in several waves. These waves out of Africa were likely spurred by challenging changes in climate or in the ecosystems that the migrating groups had been part of.

Remains found in Asia and Europe suggest that some members of the species *Homo erectus*, an archaic and extinct branch of the human lineage, wandered out of Africa well before *Homo sapiens* even existed. However, by far the largest and most significant (for us) waves of migration—those of anatomically and behaviorally modern humans or *Homo sapiens sapiens*—came only about 60–50,000 years ago.

Scholars disagree about the details of what happened when newly outbound populations met up with groups that had emigrated earlier. Those who left first may have been decimated by diseases newer population waves brought, or driven out or otherwise replaced by the newcomers (this viewpoint is known as the **replacement hypothesis**). They also or otherwise may have been absorbed or assimilated into newer populations (this viewpoint is known as the **assimilation hypothesis**). Which hypothesis will better stand the test of time, or whether the two may be productively combined, remains to be seen. The popularity of one or the other changes regularly, as new data are logged and analyzed. In some cases, genetic data and fossil data are not in accord.

The Neanderthal Situation

An example of this changing opinion has been in regard to the status of our near relatives, popularly known as the Neanderthals. This population occupied Europe and some of Western Asia (including what is now Israel) from about 130,000 to about 30,000 years ago. Some scholars have argued, based on archaeological remains, that Neanderthals are a subspecies of *Homo sapiens*—one that perhaps wandered out of Africa prior to our subspecies, *Homo sapiens sapiens*. Neanderthals then diverged from their African cousins. Without fully developed cultural capacity, they had to evolve physically to cope with their new environments, and this led to their characteristic robustness. Later, when humans fully capable of culture arrived, the Neanderthal version was assimilated back into the population. Other scholars, based on genetic evidence, have argued that Neanderthals simply were replaced when modern humans entered their habitat and pushed them from their niche. The first camp of scholars labels Neanderthals *Homo sapiens neanderthalensis*. The second calls them *Homo neanderthalensis*, denying them our species' appellation.

Those who leave out 'sapiens' thereby cast Neanderthals as what lay people might term non-human in the sense that we have defined humanity—people with full-blown cultural capacities, the sapient or wise ones. Conversely, the inclusion of 'sapiens' in their name represents them more as people like us. There is in fact archaeological evidence to show that Neanderthals did have at least a rudimentary form of culture; for instance, some did bury objects such as flowers with their dead, suggesting that they were indeed, at least to some degree, meaning-making beings. By marking them out as *Homo sapiens neanderthalensis* rather than *sapiens sapiens,* they are cast specifically as a human subspecies, or what biologists also call a 'race' (although they no longer use this term in relation to humans).

Technically speaking, a **race** is a clearly differentiated sub-group within a species. It has features that regularly co-occur. In Neanderthals, as compared to *Homo sapiens sapiens*, those included a very heavy and robust bone structure (supporting much greater strength), heads that protruded in the back, larger brow ridges and protruding noses, and smaller (or virtually no) chins. Importantly, these and several other skeletal features co-occurred with regularity: they were concordant, or had **concordance**.

Recent genetic evidence confirms that many contemporary humans have some Neanderthal genes. While no humans living in sub-Saharan Africa do so, traces of the Neanderthal genome can be found in European and Asian populations. This is, for some, concrete evidence of interbreeding, which does suggest that the appellation *Homo sapiens neanderthalensis* is correct. For others, it is simply evidence of a common ancestor, and is understood much in the same way that we understand our genetic overlap with, for instance, the chimpanzee.

Multiple Origins?

Here I should note that a minority-held variation on the assimilation theme does exist that contends that, all around the world, *Homo* populations—even the first, most archaic groups to make migratory journeys—continued to evolve separately and, more importantly, they evolved convergently, or in the same direction. That is, despite original differences, all evolutionary paths eventually came together, ending up with humans of our kind. This is perhaps most commonly called the **multiple origins** model; it has also been deemed the 'regional continuity' and the 'multiregional' model. While this perspective might be lumped here with the assimilation hypothesis, because it does allow for gene flow, it actually is distinctly different: its adherents believe that subgroups could and would have evolved in the same direction independently. Again, it is a minority position.

The Invasive Human Species

Whatever the details may have been, the larger point is that once humans could walk, some groups left Africa by foot, and this happened several times over. Among the last and by far largest waves, most scholars think, were anatomically and behaviorally modern humans or *Homo sapiens sapiens*. Whether with single or multiple regional origins, by about 10,000 years ago, our kind of humans had populated nearly every habitable area of the globe. Also, by that time they had made their mark through niche construction processes or species-environment interaction, possibly causing the extinction of numerous species of easily hunted members of the animal kingdom. Non-anthropogenic climate changes about 10,000 years ago contributed also—and many say more so—to numerous extinctions; we will return to these events later. Now, we examine changes within the human species itself.

Early Ideas about Human Origins

Only in the past few decades has our access to, and knowledge about how to read, human fossil remains allowed us to paint the picture that I have just outlined. Prior to recent scientific discoveries, people around the world were left to guess at the origins of the varied human populations with whom they were increasingly coming into contact. People put forth many theories, but they were basically of two types—and both tended to reinforce the broader values or goals of the group putting them forward.

One type of theory consisted of Adam-and-Eve type stories of human monogenesis, whereby all humans can track their ancestry to one common origin. Some theories of this type were accompanied by ideas holding that certain sub-groups devolved or degenerated while others progressed to a more perfect state (naturally, the storyteller belonged to the supposedly more advanced group). Most people who subscribed to this viewpoint held that factors such as climate or environment had an impact on human morphology (physical form) and character— that people living where it is very hot, for example, grow lazy, and so forth. Most made no reference to the idea of heredity; they focused instead on living conditions. Eventually, when Charles Darwin's ideas about evolution were brought from the biological sciences into the social sciences, scholars introduced the notion of developmental stages, theorizing that humans began in a savage, animal state, progressed to barbarism (at which point they could at least talk), and eventually could reach the most advanced state of so-called civilization (Langness 2005, Chapters 1–2).

The second type of stories created to explain human origins are polygenesis or multiple origins theories. Although some of these did eventually take on a developmental aspect, including for instance the stages of savage-barbarism-civilization, most were static; they did not include the option of change. In these theories, the various versions of humanity came into being separately. One story tells of a creator god or spirit who made us from clay and explains that color variation among groups represents how long a given group's clay forbearer was left in the fire.

Contact between very different groups of humans did not happen very frequently until the invention of transportation modes that supported it. Certainly, merchant caravans passed through many regions on foot, by camel or horse, or by wagon, and merchants did see variation first hand. However, the physical variation they encountered did not have sharp dividing lines. Rather, it turned up gradually as they passed over the land, with, for instance, hair texture grading slowly from curlier to straighter on average. Still, in hindsight, and when considering concentrations of people in major population centers, some differences would be noticed.

Which were deemed of interest and what was made of them depended on who was doing the looking and interpreting, and why.

Classifying Human Subsets

One early way differences between human groups were conceptualized was by fitting them into pre-existing conceptual models of personality or character. This is exactly what Carl Linnaeus did when he penned his original taxonomy, *Systema Naturae* (1758). As Stephen Jay Gould (1994) reminds us when describing the evolution of racial taxonomy that I will here summarize, Linnaeus was the Swedish botanist, zoologist, and physician infamous for classifying the plants and animals of the known world as part of the scientific revolution's effort at knowledge expansion. When

Figure 5.2 Four seasons, four elements, and four characters Illustration from twelfth century version of Isidore of Seville's sixth century De Natura Rerum (On the Nature of Things). Linnaeus built upon this legacy of four-fold thinking in his race typology.

it came time to classify humankind, Linnaeus decided to break us down into four groups. The primacy of four as an ordering scheme can be traced at least to Greek medicine, also known as the humoral system.

The ancient Greek or Galenic humoral system teaches that the body contains four liquids, or humors (blood, phlegm, yellow bile, and black bile), and that each is associated with a "complexion" or temperament: sanguine (cheerful), phlegmatic (lethargic or unemotional), bilious or choleric (quick-tempered), and melancholic (sad). It made associations between humors and basic elements—wind or air, water, fire, and earth—as well as with the four seasons. As long as the four humors are in balance, a person is healthy. Imbalance can be treated by removing excesses, as by purging or vomiting, or by correcting for deficiencies that can be made up through special diets (Sobo and Loustaunau 2010, 84).

As Gould tells it, Linnaeus extended the scheme to entire populations. After mapping the human race upon the known cardinal directions north, south, east, and west, he linked each with a humor and temperament. Linnaeus came up with four races that we might today term American, European, Asian, and African. He also matched them with the colors dictated by the four-fold Galenic scheme: red, white, yellow, and black (see Figure 5.2).

Despite the value placed on divisions by four, at the time Linnaeus wrote there was widespread belief among his contemporaries in the unity of the human species and an understanding that each individual contained all aspects. Linnaeus believed the differences he charted graded gently into one another, and his writing does not actually include ranking of races.

The focus on temperament in Linnaeus's scheme may also strike us as strange, conditioned as we are to look to skin color as the key to race. Cross-culturally, however, it is not bizarre at all. Many groups used such heuristics. The Lakota people of what are now the Dakotas, for instance, are said to have called white people *wasicu*. This means, literally, 'fat takers,' or 'eaters of fat'; it also means 'non-native.' It refers to the non-native's disregard for preserving natural resources for future generations; it refers to greed. In using the term to describe white people encountered, the Lakota distinguished whites not by the color of their skin but by their behavior.

Back in Europe, Linnaeus's fourfold scheme was widely accepted; after all, it fit nicely into existing ways of understanding and classifying the world. However, it was not long before someone challenged him, ostensibly on the basis of—as all good science is meant to be—observation. The challenger was Johann Friedrich Blumenbach, who in 1795 introduced the five-race schematic that will be more familiar to readers today.

Turning away from temperament, Blumenbach focused on what people looked like. He added the color brown to the four-fold white, red, yellow, and black framework. He made another change, too. Before I describe it I have to note that Blumenbach believed in the unity of humankind; he ascribed differences between human groups to climate and habitat. Nonetheless, he proclaimed people from Mount Caucasus in Georgia (a nation located at the junction of Europe and Asia, on the Black Sea) to be closest in appearance to the created ideal or original Adam and Eve, and thus justified naming his central race after them (that is, Caucasian). While geography still was important, Blumenbach's scheme made physical appearance fundamental.

Biological Race

Its origins had been laid, but a fully fledged ideology of race, complete with rankings and a biological (rather than environmental) foundation, did not gain in popularity among Europeans until the age of discovery segued into the age of conquest, and more Europeans wished to enslave or colonize more non-Europeans—particularly (but not only) those living in Africa and the Americas. At this time, the idea that people were born into fixed social stations and that some naturally could dominate others was being displaced in parts of Europe and in the emerging United States through a series of social and political revolutions. In contexts now favoring liberty and equality, those in support of slavery or colonization needed a strong justification—and they found it in the hierarchical construction of race that was Blumenbach's legacy.

What better way to justify unequal social arrangements than to claim that it is natural for a biologically more intelligent and civilized human group to enslave, colonize, or otherwise oppress another that is biologically dumber and less civilized? The popular construction of race helps people do this because of the role biology was given. People longing to justify certain socio-political arrangements came to assume that there are categorical, biologically (now genetically) based differences between broad subgroups of humanity—and that subgroups can be divided based on visually identifiable traits such as skin color.

By contrast, as a scientific rather than popular term, 'race' means 'subspecies.' It is used to refer to subspecies in many kinds of animals—but not in modern humans, because, from the perspective of biologists, no human subspecies presently exist.

Racialization

Looking at people and fitting them into subgroups based on how their bodies look (and perhaps sound) is called **racialization**. The simple classification of people according to an erroneous notion of race is distinct from

racism, which includes an evaluative component. In racism, not only are people classed by so-called race, but also the races are rated and ranked (see Fluehr-Lobban 2006, 2–4).

Everyone today who has grown up with the idea that human races exist—everyone enculturated into the belief that there are separate races, and that which race one 'belongs to' matters—will practice racialization. They will see race in people, and they will use race to organize and explain aspects of their world. With racialization, race becomes a key descriptor of every human being. This is why doctors, police officers, and others who must quickly describe patients, suspects, or clients do so by mentioning race—often prior to other descriptors ("The suspect is a white female of average height"; "You know Don—he's the black fellow who sits in the middle").

Can racialization exist without racism? Yes. We can acknowledge and appreciate difference without ranking it. This book characterizes human biocultural diversity and pays tribute to human variation. Without variation, a species may be more prone to extinction or decline, a fact demonstrated in examples throughout the book.

However, even the most neutrally deployed racialization still has a fault. That fault lies in the fact that the construct of race, upon which racialization is built, has no biological reality. That is, when we look at the biology of the large so-called races, and when we examine their genetics, *we see more difference within groups than between them*. Biological races simply do not exist.

Nonconcordance

One major challenge to the race construct comes from genetic science, which, again, has repeatedly found more genetic variation within the so-called races than between them. Whereas race has been selectively defined based on a very few of the most visible physical characteristics, human variation actually consists of all kinds of visible and invisible traits and related genes: we may or may not have the capacity to digest lactose as adults; we may or may not have lots of the intestinal transmembrane conductance regulators associated with vulnerability to cholera (see Chapter 8); we may or may not be genetically programmed for male pattern baldness; our earlobes may be long or short; our breasts small or large; chests hairy or not; cheeks dimpled or not; ear wax crumbly or moist; fingerprints loopy, arched, or whorly. We vary by myriad traits whose distribution across humanity does not overlap with race as the lay public knows it (see Diamond 1994). You may well be less genetically like a person you consider to belong to your own 'race' than you are to someone in another 'race' entirely. According to Richard

Lewontin's calculations, an astonishing 80–85 percent of human variation occurs within so-called races; a mere 10–15 percent exists between these groups (Lewontin 1972, 2).

This fact seems shocking even now, more than forty years later, because of our culture's deeply held belief in race and our reliance on visual cues to mark it. The statistics just given describe genetic differences, however, not just easily seen ones: many of our genes store information used by our bodies in making proteins expressed under the skin and internally. As many communities have asserted in support of our common humanity, we all have red blood. We also all have lungs, bones, a liver, and so on. In addition, features that make the human species unique—our brain-based capacities for logical and creative thought and for empathy and the ability to cooperate and form communities—are longstanding and account for much more of our humanity than those features that are evolutionarily newer and geographically varied.

Moreover, the genes containing information used in building those visible traits we most often associate with race are not actually distributed in race-like clumps. They do not occur only in certain so-called racial groups. More important, they do not co-occur, or occur together, in any predictable way. Rather than concordance (when one gene set, gene, or allele is a predictor for or index to the fact that another particular gene set, gene, or allele will co-occur), so-called racial traits exhibit **nonconcordance**.

This may seem counterintuitive. We experience the world through a racialized lens. Yet our belief that races are out there is similar to the pre-Copernican belief that the world is flat. It looks flat from my window even today. However, in truth, it is shaped like a marble. Also, in truth, racialized traits do not necessarily co-occur. Many are inherited independent of or nonconcordant with other racialized traits.

Traits that do co-occur usually do so in subsets of people within a larger 'racial' category—subsets having to do with specific ancestry. So, for example, we do find certain patterns of concordance among people whose ancestors come from a particular village in England, or a particular village in Equatorial Guinea, or a particular village in China. However, these concordant traits are not found concordantly in all people from England, or Equatorial Guinea, or China; they definitely are not seen among all whites, or blacks, or Asians.

Moreover, so-called index traits like skin color are not really distributed in a way that overlaps entirely with the racial groups we have invented. While it is the case that many people living in sub-Saharan Africa have darkly pigmented skin, so do some people indigenous to other areas: certain Indians, Melanesians such as the people of Fiji and

Papua New Guinea, and Australian aboriginals are just some examples. Likewise, some Africans have eyes in the so-called Asian shape, with epicanthic folds, although they are not Asian.

Another way to think about nonconcordance is to consider the differences between, say, Haitians, Sudanese, and Ugandans. All may be racialized as 'black' but these groups are very different, historically as well as physically (Relethford 2000; Bruner and Manzi 2004). Build varies, as does height: African peoples can be tall, as among the Maasai of Kenya and northern Tanzania, or they can be short, as among the Mbuti who live in the Congo region. Likewise, facial structure varies: it is longer and narrower to the north, and shorter, broader, and flatter to the south. In terms of skin color, which is taken so often as the ultimate marker of race, we actually find more pigmentary variation between people living within sub-Saharan Africa than in any other geographic region in the world (see Figure 5.3).

The logic of grouping all African people together starts to seem nonsensical once we consider the wide variation of traits within Africa. Likewise, Peruvians, Costa Ricans, Mexicans living in Chiapas, and Mexicans living in San Diego are very different in average height, facial structure, skin pigmentation, and so forth, putting the lie to any idealized lumping of them into a broad, racialized group. Or for another example, there are the long-term English, French, Irish, Polish, and Italian populations. All may be classed as white today, but there are differences among them large enough to have at one time caused some to be labeled 'non-white' (see Fluehr-Lobban 2006, 180–183).

Belief in 'Race' Fosters Clinical Mistakes

Some subsets of populations do bear being defined biologically, including genetically, not least because of the clinical implications that ancestry may have. People whose ancestors are from region X may be more likely to suffer from condition Y or be more susceptible to disease Z. However, in making these distinctions we must not confuse ancestry with race. It is crucial not to wash out difference by throwing people of varied ancestry (for example, Basques and Catalans, both of whom live in what is now Spain, or people from Spain in general and people from Italy, Albania, and the United Kingdom) into the same racialized bucket. Such over-generalizations have little real utility in physical terms and in fact can be detrimental medically if particular propensities are overlooked because they don't fit a patient's 'racial profile.'

Richard Garcia (2003) has spoken out against racial profiling by clinicians. Race is a constant descriptor in medical records, as if it can tell us

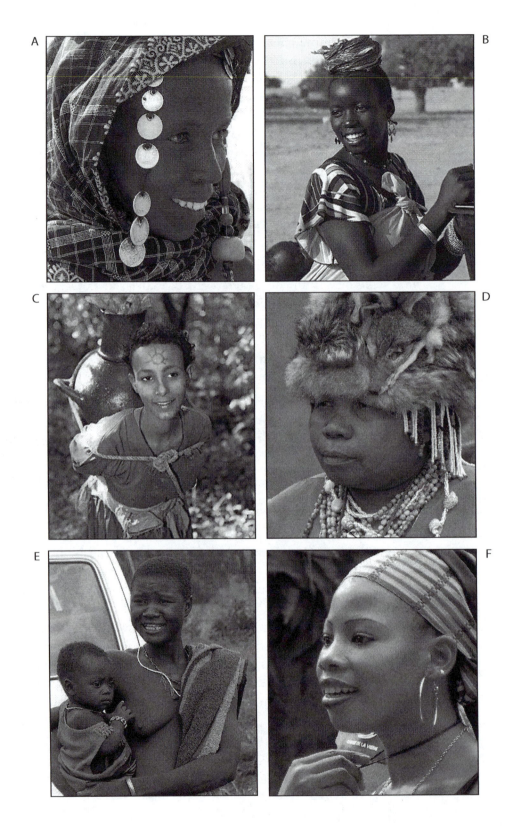

something important about a patient. Instead, it seems to cause otherwise smart people to fall back into stereotypic thought patterns. For instance, Garcia tells of a black teenager with a cyst in his head so large that it had "squashed his brain against his skull." Emergency room staff initially wrote off this youth as simply an intoxicated gang member. In another example, Garcia's childhood friend, Lela, suffered from cystic fibrosis (CF) for much of her early life without a proper diagnosis because of an assumption that people in her racial category ('black') did not get this disease. As Garcia tells it, "Only when she was 8 did a radiologist, who had never seen her face to face, notice her chest X-ray and ask, 'Who's the kid with CF?'" (2003, B15). (For more on CF, see Chapter 8.)

Things are further complicated by the high frequency of 'mixed race' children, such as Garcia's own blonde, fair-skinned daughter, whose mother is 'black.' Any doctor who simply eyeballs his child may inadvertently and erroneously cross off of the 'possible diagnoses' list any disease 'known' to predominate in 'non-white races.'

Forensic Anthropology and 'Race'

While most scholars and scientists have long discarded racial categorizing in favor of finer-grained inquiries into geographic ancestry, it remains in use among forensic anthropologists who do skeletal identification for law-enforcement agencies, even though those whose work concerns blood groups tend to reject it. Why? As forensic anthropologist George W. Gill explains, while race may not be useful for analyzing blood, it surely is for analyzing bones (2000).

Certain skeletal traits, says Gill, do prove reliable in "assessing ancestry," as forensic anthropologists term it. Specifically, a number of individual methods, such as those entailing midfacial measurements or femur (thigh bone) traits, have an accuracy rate of over 80 percent on their own; used in combination, these methods can predict race with near-certainty. This is a fact.

However there are two problems here. One is that living people generally use skin color as the primary or key index of race; while they can see some effects of skeletal structure (for instance on the face), they do not look first at a person's skeleton—or even his or her facial

Figure 5.3 (*Opposite*): (A) Young Peulh, Mali. *Copyright Paul Smith; used by permission.* (B) Bozo woman at the village well, 2010. *Eskil Fredriksson.* (C) Gojam woman, Ethiopia, 2008. *Daniel Cherie.* (D) Traditional healer, Lesotho, 2009. *Aaron Frick.* (E) Young mother in Tamberma, Togo, 2005. *Copyright Paul Smith; used by permission.* (F) Woman in Bamako, Mali. *Copyright Paul Smith; used by permission.*

structure—to decide what race box to plop that person into. A person with a 'white skeleton' may actually have been very dark-skinned, and so on. Another problem is that the racial classification scheme is, biologically, bogus. As Gill explains:

> The "reality of race" … depends more on the definition of reality than on the definition of race. If we choose to accept the system of racial taxonomy that physical anthropologists have traditionally established— major races: black, white, etc.—then one can classify human skeletons within it just as well as one can living humans.… I have been able to prove to myself over the years, in actual legal cases, that I am more accurate at assessing race from skeletal remains than from looking at living people standing before me. So those of us in forensic anthropology know that the skeleton reflects race, whether "real" or not. (Gill 2000)

In effect, then, 'racing' skeletal remains can work. We can determine, at a coarse level, whether a person was black, white, Asian, and so on. However, such classing is based on convention, not biological fact. Because the expectation for it has been institutionalized, job security for forensic anthropologists can hinge on fulfilling requests to 'race' remains. That begs the question: does it provide added value to a forensic report?

Some critics chide that racing skeletal remains is like determining whether the creature encountered in a forest glade was an elf or a hobbit or a gnome when none of those creatures really exists. Others note that 'racing' bones diverts attention from facts about a person that are likely to be more crucial to his or her identity, identification, and perhaps cause of death. For instance, someone whose remains were identified as 'black' may have in fact been a business traveler who self-identified first and foremost as Norwegian. That knowledge, if available, would throw his or her case in an entirely different light. So, while it is possible to do or provide racial classification in forensics, like racial classification in clinical medicine, it poses a problematic diversion. Just because we can do it does not mean that we always should. This is particularly so in arenas linked to the biological sciences, which squarely reject the concept as applicable in human beings.

Census-type Data

I do not contend that all racial data are useless simply because race is not a biological entity. Race exists as a cultural construct, and it has real ramifications in terms of life chances and experiences. Therefore, we do need to collect such data sometimes.

Race data can be used for tracking how well or how poorly we are doing in terms of meeting certain goals within particular institutions, for instance. If we did not track people by race, we would not know, for example, about particular problems in educational access or achievement, or inequities in healthcare. I myself have collected race data in the context of pediatric quality improvement research. I generally used self-reports—although the 'eyeball method' also has merit in that context, if only because it may better reflect how clinical professionals are likely to categorize an individual, and so it may better serve as a predictor of the reaction or treatment that those individuals are likely to get (as Dr. Garcia has already warned us).

Ethnicity

It does seem inappropriate, however, to use the term 'race' when what we are measuring is really not biological, and so there has been a slow shift toward use of the term 'ethnicity.' A full-fledged discussion of ethnicity could take us far from our path in this chapter, so here let's just note that, technically speaking, **ethnicity** is tied to notions of shared national or regional origins and shared culture—but not necessarily shared biological heritage.

In the United States, ethnicity is reflected in the government census classification distinguishing 'Hispanic or Latino' people from those 'Not Hispanic or Latino.' These options are used to supplement the question of whether a person is 'American Indian or Alaskan Native,' 'Asian,' 'Black or African American,' 'Native Hawaiian or Other Pacific Islander,' or 'White' in terms of 'racial origin.' Thus, in the United States, Hispanics who have more long-term ancestors from Europe (for example, Spain) than from, say, Peru, may class themselves as White Hispanics. Those with more long-term ancestors from the Americas might class themselves as non-White Hispanics, as may those whose long-term ancestors were mostly brought to the Americas (for example, Panama) from Africa as slaves.

Note that the U.S. census is just one of many census systems that have tried to grapple with the problem of how to subdivide a population. Because each nation has different ideas about what kind of background data are relevant in describing the population, each nation's census forms have different options. For instance a 'black' person in England could be counted by the census there as Caribbean or African; but in Guam, as in the United States, there is only one category for black individuals. The English also subdivide 'Asians,' while a person from India and a person from China would be counted in the same category in the South African census (American Anthropological Association 2011).

In many nations racial and ethnic designations used by the lay population differ from those used officially. In the United States the general public holds that there are plenty more ethnic groups than just 'Hispanic or Latino' (and its converse, 'Not'). As well, many feel that the categories are themselves much too broad, and so speak of Mexican Americans versus Cuban Americans and so on. We also speak of German Americans, Polish Americans, Indian Americans, African Americans, Japanese Americans, and the like. Note that all of these titles make an overt reference to peoples' shared ancestral origins and broad cultural heritage (including linguistic, religious, and so on). In effect, and although on their own they also can lead to dangerous stereotyping, ethnic classifications come much closer to the fine-grained groupings related to geographic ancestry that doctors and others would find useful.

Of course, ethnic categories can themselves lead to problematic generalizations. Some ethnic categories are so broad as to reveal insufficient detail about ancestry. Also, ethnicity has a very strong cultural component—ethnicities can be learned over time, or forgotten. Immigrants to California may come to identify with and pick up expressions and customs that predominate in their new-found community—expressions and customs that differ from those they might have learned had they migrated to New York City. Likewise, the regional basis for ethnicities can collapse, expand, or even be forgotten. White middle-class American ethnicity, for example, is attributed now to people whose ancestors came from all over Europe—even though important sub-group distinctions were once made among Europeans. Today, members of this ethnic group do not generally think much about their particular familial European history, or even consider themselves members of an ethnic grouping. Nevertheless, they are communally identified through commonly preferred television shows, modes of dress, and viewpoints, including views on who qualifies as a member of their group.

Geographic Clines

As Blumenbach knew so many years ago, we know today, too, that the boundaries between so-called racial groups are porous. Rather than hard dividing lines, there are blurry areas between these groups. The blur actually stems not from group-group bleed but rather from trait-trait gradients. Individual traits blur at the borders at very different rates. Moreover, some traits once thought to be confined to one group actually are found (sometimes at quite high rates) among another. Pigmentation in the skin provides a good example. Many groups of people living in India are much darker than the group of people classed in America as 'black.'

Rather than being linked to so-called race, human variation is linked to geography and to the adaptive challenges posed by particular regions. Such challenges generally do not appear and disappear as if controlled by an on-off switch; they don't simply exist or not exist. Rather, they grade in and grade out, or arise incrementally and dissipate similarly, as vegetation and air quality change when one goes up into and then descends down out of the mountains. The incrementally changing distributions of traits over geographic regions, which are related to incrementally changing challenges, are known by biologists as **geographic clines**.

The Example of Skin Pigmentation

Because they are geographic, clines are best demonstrated on maps. Maps of skin color distribution, for instance, show that pigmentation or darkening of the skin with melanin is heavier in areas where people are more exposed to a certain form of ultraviolet radiation (UVB, or B-range ultraviolet radiation), and that it gets lighter and lighter the farther away from those regions populations get. Why? As Nina Jablonski and George Chaplin explain it (American Anthropological Association 2007; see also Lenkeit 2009), melanin keeps UVB from penetrating through to the bottom layers of our skin. Think of it as a barrier to UVB. Still, why would blocking UVB give a population a selective advantage?

Some UVB is necessary for the creation, in the human body, of vitamin D, which the skin actually creates in response when exposed to it. Vitamin D helps the body absorb calcium, and sunlight-generated vitamin D is better at doing this than the dietary variety, which, in any case, is not available in all environments (fish oil is a great source). Calcium helps build strong bones and teeth. However, too much vitamin D can poison us, and this is more likely to happen to those who live in sunny climates but without much melanin in their skin. Darker skin in such regions serves an adaptive function. So does body-covering clothing, but that of course is a cultural adaptation that didn't exist prior to the emergence of culture.

While too much UVB can lead the skin to overproduce vitamin D, it can, conversely, lead to the destruction of folate or folic acid. Folate is essential for successful reproduction, and so those who were protected against losing it were likely to have more healthy offspring, increasing the frequency of the genes that code for more melanin production in the next generation's gene pool.

Skin pigmentation patterns around the globe—at least prior to mass migrations and jet travel—reflect the importance in attaining a balance of melanin that will promote ample calcium absorption and protect folate stores given the degree of UVB exposure. Lighter skin, with less pigmentation, was naturally selected for in populations that ended up living in areas with low UVB exposure, such as those living in Scandinavia. Conversely, populations living in certain parts of Africa and South Asia have very dark skin. Populations graded from darkness to lightness in geographically clinal patterns, just as UVB penetration of the atmosphere appears as a gradient when mapped.

An apparent exception concerns people living in regions of Canada and Alaska, where sunlight is very scarce at certain times of the year—but who are relatively dark in color. The explanation for this has to do with their diet, which is high in vitamin-D rich foods, including seal and walrus; the high degree of reflected sunlight they are subjected to when it bounces off ice and snow; and their relatively recent arrival in that region (evolution typically takes a long time).

Other Good Examples

Other traits that are good candidates for demonstrating geographic clines include head size, ABO blood group, the body's ability to cope with large quantities of alcohol, and whether or not one carries the sickle cell allele or can digest milk sugar as an adult. Recall that the latter two are found, respectively, in geographic regions where there is malaria, and where there are sheep or other dairy animals. The ability to metabolize or break down and assimilate, without problems, large quantities of alcohol seems to be found more frequently where populations lived without a reliably safe or pure water supply and with access to grains that might be fermented, such as in the densely populated agricultural areas of Britain (Ridley 1999, 191). Blood type distributions may be linked to a history of exposure to particular epidemic diseases; people with A or B types appear to be protected from particular diarrheal diseases (and the AB type more so), many of which swept through the areas in which this blood type is most common. Type O seems to confer some advantage over malaria, and is thus more commonly found in areas where that disease is historically endemic (Ridley 1999, 140–144).

Even body build is clinally distributed. For instance, a tall, slender build is of greater advantage in regions that are very hot, such as in northeastern Africa. There, having more body area allows for more sweating and thus better thermoregulation. Stocky builds, which help to conserve body heat, are better for colder areas like the Arctic—and not surprisingly are found there (McElroy and Townsend 2004, 81–82).

Bad Examples

We should not make the error of assuming, based on this logic, that all traits are clinal. Nor should we assume that contemporary pockets of particular traits resulted from clinal pressures. Not only does widespread human migration muddy the waters; so, too, does the fact that we must always come back to—despite the above-discussed exceptions, most traits are not purely biological, let alone genetic.

Take, for example, Jamaicans, who at the 2008 Olympics won eleven medals in track and field for their tiny island nation, including six gold. In 2012, they won twelve (four gold, four silver, four bronze). This was expected based on their 2008 results—but in 2008, their success took the world by surprise.

Racialist thinking sprung into action as people tried to explain their outstanding success (and racist thinking, too, as 'really hard work' and 'dedicated, improvement-directed practice' were overlooked as possible reasons). It is true that the geography of Jamaica and other environmental factors might have led to selection, in the island, for some kind of biological advantage. As David Shenk notes, the media had found one "within hours" of the 2008 successes, calling it Jamaica's "secret weapon": the protein alpha-actinin 3, which powers muscles to contract forcefully and fast. The protein, it turned out, was linked to the ACTN3 allele—and 98 percent of all Jamaicans had it (Shenk 2011, 100–101).

Yet, as Shenk reports, "no one stopped to do the math" (2011, 101). It turns out that in the United States 80 percent of all people also have a copy of the allele. Nobody stopped to do the geographic investigations either: it turns out that Jamaica is an island with numerous microclimates and varied human habitats. Moreover, Jamaicans themselves are extremely heterogeneous genetically (let alone otherwise). Like African Americans, whose heritage is much like their own, Jamaicans do not represent a genetic island; they generally have a mix of West African, Indigenous, and European ancestry, but proportions range widely in individuals, and many have Asian heritage, too. Partly because Jamaica served as a crossroads for the slave, sugar, and rum trades, it—and its inhabitants—became very cosmopolitan. Athletic clusters, such as are found in Jamaica, "are not genetic, but systemic" (101)—they thrive in the interaction of biology and culture.

In Jamaica, people have long revered track and field—a fact highlighted by field and track athletes themselves when interviewed. In the past, great athletes left the country for others where they might better develop their careers. However, in the late 1960s, returning athlete Dennis Johnson helped the small, poor nation begin to develop an elite training program (Shenk 2011, 109–111). This, plus the intense love of running and can-do mindset fostered by Jamaican culture (who can forget the Jamaican bobsled team's 1988 performance? see Figure 5.4), much better accounts for Jamaican Olympic success than does trying to explain it as clinal, let alone 'racial.' Particularly in a context where poverty limits opportunities, so that sports achievement becomes a golden ticket worthy of self-sacrifice and devotion, Jamaica's Olympic success makes sense.

Race's Social Significance and Self-Perpetuation

The information in this chapter is not meant to suggest that race does not influence our lives. It surely does. In fact, it may in some sense be a self-serving, self-perpetuating construct—one that helps to generate genetic difference, albeit in tiny measure relative to other pressures. For example, people who seek to marry people more like themselves may be aiding the sexual selection of particular racialized traits—traits that members of a particular group find more beautiful. Indeed, some argue that this kind of selection explains the population-wide emergence and persistence of racialized traits for which we now have no satisfying adaptive explanation (for example, larger buttocks in certain female populations, more body and facial hair in certain male populations; see Diamond 1994).

On a more sinister level, racialization often gives way to racism—the prejudicial and biologically determinist belief that some races are better than or superior to others—a belief often grounded in reductionist thought. Such prejudice can have and has had terrible effects on populations deemed inferior as well as, some would argue, on the moral health of those who are racist. For populations deemed inferior, racism can limit access to institutions and mechanisms—such as education, healthcare, or the right to own property—that would otherwise allow for the optimization of developmental potential. Racism even can lead to a group's direct physical persecution, for example lynching, enslavement, or even genocide.

The effects of racial thinking can reach the genome, too; in promoting sexual selection it can lead to small population differences that were not there to start with. For instance, Tay-Sachs is a lethal metabolic disease found more frequently in Jewish people of specifically Eastern European or Ashkenazic heritage (**metabolism** is the biochemical process of breaking down and repurposing food components for bodily use). The gene

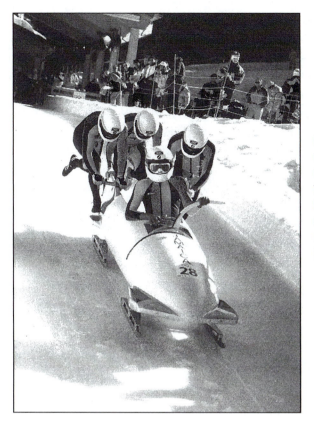

Figure 5.4 Jamaican atheletes Ricky Mcinstosh, Nelson Stokes, Dudley Stokes (*driver*) and Michael White at the start of the four-man bobsled race at the La Plagne Winter Olympics, France, 1992. *Photo by Leo Campbell, courtesy of N. Christian Stokes.*

sequence entailed in Tay-Sachs may have provided a heterozygous advantage against tuberculosis, or TB, which would have been a grave threat to survival of ghettoized Jews, such as those living in Nazi-occupied Poland (McElroy and Townsend 2004, 90). In this, Tay-Sachs carriers are much like those who carry the gene for sickle cell disease: while the latter have higher survival odds in the face of malaria, the former have higher survival odds in the case of TB. Like the gene for sickle cell, the gene sequence that causes Tay-Sachs when homozygous was, originally (before the last waves of Jewish migration), regionally and clinally located. The difference of note here is that overtly racist housing regulations underwrote the development of this human variation.

One Race Not Many

We will return to the contemporary ill effects of a belief in race later, after we build our understanding of the basis for present-day social structures. This chapter has set us up to do that successfully because race should no longer divert us.

We have looked under the surface of the race construct to see what it attempts to summarize about human biology. We have learned how poorly fitted it is to its categorizing task due to the many variations in human ancestral habitats—including some human-made—and the resulting broad range of diverse and variously overlapping adaptations that different gene pools carry. These pools—we humans—are not so easily or clearly divisible into a handful of broad racial categories. Rather, diverse human traits are best understood as distributed along continuums, many of which have very little to do with one another. Further, along a given continuum, there are so many slight gradations in a trait that deciding where one type ends and another begins can be a futile business indeed. A more productive approach to human diversity is one that uses a holistic lens. The next chapter brings us closer to that goal by exploring how humans—both those who left and those who stayed in Africa—made their living once they had culture.

Socio-political and Economic Factors

In Part I, we adopted a systems point of view toward human diversity, taking into account the multi-directional interactions between various system components—interactions that sometimes led to the emergence of something—or some trait—entirely new. We learned how advantageous variation within a population can be to group survival. From a systems perspective, many of the variations now existing between human populations are the legacy of human-environment dynamics in which particular traits, under particular circumstances, provided an adaptive advantage.

Although we did not ignore humanly created environmental factors in Part I, in Part II they come to the fore. Part II focuses particularly on the consequences for human well-being of how we subsist or make a living, and how we organize ourselves socially and politically. Ideas introduced in Part I about synergy take on increased importance as we explore the truly interactional, co-creational nature of our biologies and our cultures.

Cultures include pragmatic knowledge, such as how to raise certain crops, and agreements on how people should organize themselves in terms of power hierarchies and for the exchange of goods. Part II focuses on how the socio-political and economic structures humans create, along with habitat features, relate to the kind of human diversity already described.

As an added feature, Part II highlights health as an index of how well—or how poorly—a group is adapted to the circumstances in which it lives.

Chapter by Chapter Overview

Part II begins with Chapter 6's examination of health-related correlates of the forager (hunter-gatherer) way of life. After exploring the ramifications of foraging for nutritional status and other aspects of human health, Chapter 6 describes the selective advantages of breastfeeding and of co-sleeping with infants and children, and explains how and why these evolutionarily appropriate patterns might be bypassed in contemporary circumstances. In this way, we begin to ask whether some 'modern' practices may in fact make little if any evolutionary sense.

Except for scavenging, foraging is the most basic way of making a living. A more intense method is settled agriculture. After characterizing the subsistence continuum and intensification process implied here, Chapter 7 describes the impact that a shift from foraging to agriculture can have for socio-political and economic organization. After asking how and why domestication happened where and when it did, we investigate the ways that, through complex feedback loops, sedentism, surplus accumulation, specialization, and social stratification evolved—and how they led (as they continue to lead) to poorer health for large portions of agricultural populations. Agriculture-related health effects seen in the archaeological record, including the skeletal record, are examined; and to provide a point of contrast, we also explore the biocultural effects of industrialized agriculture—the kind we practice today.

In the first two chapters of this section, then, we move from our initial explorations of the adaptations that fostered humanity's emergence, via natural selection, to those adaptations made by humanity in response to the very ways of life that behaviorally modern humans had created. We examine how subsistence strategies affect not only health but also social structure and then, via social structure, affect health again. This kind of multiplicative feedback can and does happen because human systems work much like complex adaptive systems, in which everything is connected.

With Chapter 8, we begin to bring in a few new theoretical frames to further our quest to comprehend the bioculturally interactive nature of human variation. After instruction in the basic language of epidemiology, and a review of the role of the human immune system in the body-environment relationship, readers learn to use the disease ecology perspective to explain how the agricultural lifestyle contributed to the spread of infectious diseases. This lesson is extended to show how humanly created changes in our environment today have led to drug resistance (or increased staying power) in certain strains of infectious disease, while lessening the severity of others. As well, our own bodies have adapted, in certain ways, in certain populations, to the stresses brought on by humanly fostered epidemics, so that some populations are more (or less) prone to certain disease than others.

A second framework emphasized in Part II focuses on what is termed the 'political economy' (put loosely, contemporary global capitalism and its antecedents). The biocultural diversity fostered by, for instance, the class system is certainly humanly created , and not all for the good. In Chapter 9, readers learn to use the political economy perspective to gain insight into the differentially distributed impact of social stratification on human populations, particularly as seen in health indicators and outcomes. But Part II does not end on a negative note; readers learn about the social justice approach as one way to help combat the humanly induced biocultural variation reflected in health inequity statistics.

Central Lessons of Part II

By the end of Part II of the book, readers will have sharpened their understanding of the relationship between biology and culture entailed in the assertion that we are biocultural beings. They will be able to identify and explain how the social structures—which humans, by dint of their evolved cultural capacities, have created—can themselves foster and maintain population-level biocultural differences. These differences range from naturally selected for genetically based resistance to particular diseases in some populations, to developmentally derived cognitive deficits related to toxic exposures or nutrient shortages in others, to context-dependent enhancements to running speeds and other athletic abilities in still other groups.

As well as being able to identify and explain such differences, readers will have gained practice in seeing connections. Readers will have grown more skilled in seeing how the ways that a society engages with the environment (their subsistence mode) will have knock-on effects, and in asking how populations that seem to be worlds apart geographically, socially, politically, and economically, let alone bioculturally, may in fact be intimately implicated in each other's life chances and challenges.

Foraging: A Human Baseline

This chapter prepares you to:

- Characterize the typical foraging or hunter-gatherer subsistence strategy

- Explain the ramifications of foraging for nutritional status and other aspects of human health

- Describe the selective advantages of breastfeeding and explain how and why the evolutionarily appropriate pattern might be bypassed

- Describe the selective advantages of co-sleeping and explain how and why the evolutionarily appropriate pattern might be bypassed

In previous chapters, we learned about human adaptation, including the kinds of adaptation that led to the evolution of behaviorally modern human beings—people like us. With the emergence of our capacity for culture, we spread out both within and beyond Africa, eventually colonizing the entire planet.

Our original **subsistence strategy**—our approach to the task of extracting food—was to hunt and gather, also known as **foraging**. We collected all kinds of plants, insects, and animals to eat, and infants and toddlers suckled at the breast.

Because different regions have different kinds of natural resources, one of our first major tasks upon moving to any new environment entailed figuring out how to adapt our foraging skills so that we might extract ample amounts of food and water locally. In some geographic locations, exploiting resources that differed quite substantially from those we had previously evolved to depend upon also meant we had to develop new tastes. There were various fruits, leaves, herbs, grains, legumes, fish, crustaceans, insects, reptiles, birds, and animals wherever we went, and different subsistence traditions resulted.

No matter how our diets differed from place to place, we remained foragers for thousands and thousands of years. Our current dependence on industrial agriculture is a very recent adaptation. Even the time passed since the emergence of basic agriculture about 10,000 years ago accounts for just a drop in the bucket of our biocultural history.

Nutrition Basics

Human beings are part of an environment in which the sun plays a key role as an energy source. When we eat, we capture and repurpose energy originally sent by the sun. The sun shines down on the plants, the plants grow with the energy of the sun, and we eat the plants (or the animals that eat them first). The idea of a food chain, which places the animals that are not prey for others at the very top, and uneaten organisms into the soil upon death, reflects this process of conversion.

Our present nutritional needs are the result of a long-term process of genetic evolution whereby our primordial ancestors were transformed from creatures lacking even mouths (let alone brains, stomachs, kidneys, intestines, and so forth) into complex animal organisms (us). Our rudimentary ancestors simply absorbed the sun's energy and certain environmentally available elements or compounds that washed over them; but as humans, we must actively eat and we must drink.

Our *Homo sapiens sapiens* ancestors were tall; we know this from skeletal data. They also were lean, in part due to a vigorous physical lifestyle. Their nourishment involved—as does ours—the intake of three fundamental substances: protein, carbohydrate, and fat. Protein is, in effect, the material substance from which we build our bodies. It is in turn created from amino acids, found in diverse combinations in various foods items. Complete protein sources such as meat do exist, but we also and sometimes only get protein from juxtaposing foods with complementary arrays of amino acids. Some examples are corn and beans, whole grain cereal and milk, rice and peas, and peanut butter and whole grain bread.

While protein provides matter in effect, energy—the energy to build body parts from that matter, or to use it, for instance, in running or picking onions—comes from carbohydrates (sugars, starches). Fat serves as an energy store, and for insulation; it also is essential for proper functioning of the nervous system.

The three key classes of nutriments—protein, carbohydrate, and fat—are called **macronutrients**. They are pulled from various foods through the digestive process and then put back together in different ways to structure our bodies or serve as fuel when needed. To correctly process the macronutrients we also need trace amounts of what are therefore termed **micronutrients**: vitamins and minerals without which the biochemical processes entailed in keeping us going just cannot happen. Likewise, water is essential to human survival.

Prior to the days of the supermarket and trans- and intercontinental shipping (and even still today in many regions of the world) macronutrient and micronutrient sources varied depending on what was geographically available. For people who lived near rivers or bodies of water, fish and crustaceans could provide many nutrients; not so for those in the desert. Likewise, certain fruits that were abundant in central Africa simply were not available to those living in the North.

Further, many of the food items we think of today as essential to a good diet were unknown in older times. Bread as we know it, for instance, did not exist until well after the invention of agriculture. Milk, too, was not standard fare. Indeed, to digest such things properly, populations exposed to them adapted genetically, over time. Given that, what was our diet really like prior to farming, and how did we keep ourselves fed?

The Foraged Diet

Our human ancestors, like most foragers today, were relatively tall and lean. We know this from skeletal data and by making ethnographic extrapolations. We also know that their lifestyle demanded plenty of physical activity. Therefore, as Eaton, Eaton, and Konnor argue, they "must have existed within a high energy throughput metabolic environment characterized by both greater caloric output and greater caloric intake than is now the rule" (1997, 208). In other words, they both used (metabolized) and ate lots of food.

What kind of food did they eat? One common misconception about our early diet is that it was heavy on meat; this is reflected in the misnomer 'hunter-gatherer,' and it is why some scholars prefer the term 'forager.' In reality, most of the food that hunter-gatherers or foragers eat generally comes from plants (see figure 6.1). When I say 'plants,' I do not mean cereal crops such as oats, rice, maize, barley, or wheat—those are more commonly cultivated than foraged. Rather, I am talking about roots, fruits, nuts, legumes, leaves, and other non-cereal plant items. These make up about two-thirds of the average forager subsistence base.

Moreover, we often foraged for insects and arachnids; this would have been quite common in the past, and even today many people still gather and eat them. Bugs did, and do, make up a substantial portion of

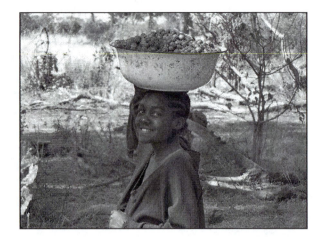

Figure 6.1 Gathering food in the Okavango Delta, Botswana, 2004. *Photo by w:User:DDD DDD. Public domain via Wikimedia Commons.*

the protein that many people get in their diet. Those who quantify these things argue that a bug rather than animal based diet could even save the planet. Bugs are much more efficient at converting what they eat into 'meat' than cattle, and their 'meat' is highly nutritious—with grasshopper, for instance, providing three times the protein of beef, ounce for ounce. Plus, bugs have a significantly smaller footprint on the environment than livestock; they are "natural recyclers" and actually thrive under low-cost conditions. Bugs can be quite tasty, too (Goodyear 2011).

If the thought of eating grasshoppers and their ilk elicits an ethnocentric 'ugh,' even in light of sustainability issues, consider the following. Many Americans and Europeans, for example, happily eat what to others is putrid, moldy, and rotten: cheese, which is indeed made by decaying milk, sometimes with added bacteria, too. Furthermore, Americans who may not like the idea of eating bugs eat them anyhow; for instance, children's beloved peanut butter contains bugs. According to the U.S. Food and Drug Administration, peanut butter can have up to 30 insect fragments per 100 grams before it is subject to seizure or grounds for a citation. We'll talk more about the industrial food system in the next chapter.

What of the quality of the forager diet? Again, we turn to Eaton, Eaton, and Konnor, who have studied the question extensively. They conclude that it is "inescapable that our preagricultural ancestors would have had an intake of most vitamins and minerals much in excess of currently recommended dietary allowances" (1997, 208). Indeed, their diet was "nutrient rich" (212). Further, in contrast to our diet, foragers' carbohydrate and fat intake was in line with today's recommendations. This is important, because of the links between health problems such as diabetes and heart disease and an overabundance in the diet of carbohydrates and fat (especially saturated fat, which is much less abundant in the typical forager diet). Any excess protein ingested by foragers generally would, by today's standards, have been offset by higher fiber and potassium intake and lower levels of sodium, as well as more physical activity.

So much for solids. What about liquids? Water, in pure form, is not always in constant or easy supply. People who lived in geographic regions without ample potable (clean, drinkable) water sources had to discover or invent ways to get good water or they would not have survived: we are ill-fitted to a water-free environment. Water can be gotten from food, of course, such as juicy melons or tuberous roots. People who raise milk-giving animals can get water from fluid milk, once they evolve the adaptive ability to digest it. Beverages that contain water but which are sterile and clean, such as beer or wine, also provide fluids.

Foragers do not, by definition, herd animals and most do not need to make alcoholic beverages—but if I may digress momentarily a broadly relevant point can be made. In communities in which large quantities of beer or wine are regularly drunk in lieu of potable water, people who are less vulnerable to over-intoxication or drunkenness have a survival advantage. Who is less vulnerable? Group members able to increase production of the enzyme that breaks down alcohol. In beer or wine drinking cultures of old, these group members had increased fitness; more of their genes were passed along into the next generation. Further, with alcoholic beverages so central to the diet, cultural rules regarding things like when, with whom, where, and how much to drink were put into practice. Therefore, people whose ancestors relied on fermented drink often have inherited—both biologically and culturally—the capacity to drink large amounts of alcohol with few or no problems. On the other hand, those whose ancestors hail from parts of Australia and North America, where there was plenty of good, clean water, and therefore no culture to speak of regarding alcohol use, may themselves be quite vulnerable to alcohol's ill effects (Ridley 1999, 191).

Drinks aside, the data clearly support the hypothesis that our early human ancestors' forager diet was different than ours in terms of the ratio of plant to animal items, and the balance of protein, carbohydrate, and fat. Geographic variation affected the specifics of this—it affected what they took into their bodies for nourishment, and how they metabolized foodstuffs. It had knock-on, secondary effects on other bodily systems, such as cardiac health—and on culture, too.

It is less patently obvious that the nutritional value of the food items themselves was different. Yet, based on current agricultural science and ethnographic data from contemporary foragers, we can assuredly say that it was.

First, most foraged food items are consumed within hours of being collected and with minimal processing. That is, they are eaten fresher and rawer than most food today. If I pick an apple from a tree and eat it directly or even one week later, I get a better deal nutritionally speaking than I get when I buy an apple (or potato, or whatnot) from the grocery store—one that's traveled on a truck across the country and had been in cold storage for a month or more prior to making its trip. In that time it will lose some freshness; it will lose some nutrients. They degrade, or seep away. A lack of attention to storage conditions, such as temperature and humidity, can make matters worse, as can infestation with pests, including molds and fungi. When nutrients break down or degrade their **bioavailability** to the eater—the degree to which the eater can extract and make use of them biologically—is negatively affected.

Cooking also can break down components in a food item. In some cases that can be good, increasing the bioavailability of certain nutrients by freeing them up, so to speak. Some convincing arguments have been made suggesting that the evolution of humans' big brains could not have occurred without the value added to certain foods through cooking. Whether the key food here was meat or yams is a matter of contention. We do know for certain, however, that cooked food (vegetarian or not) is easier on the digestive system and jaw. Those born with bodies set to put less into building big teeth and long guts and more into building the brain were no longer at a selective disadvantage once cooking had been invented.

Cooking also can leach nutrients from food or decrease their nutritional value. When we boil certain vegetables, for instance, some of their nutrients are left behind in the pot water; and the longer we boil them the more nutrients are broken down or drawn out. Some forms of cooking can add components that are simply not healthful, as when frying, or grilling on certain charcoals. Toxic compounds also can be introduced to food on contact with cookware or cups and dishes, for instance in the form of lead used in glazes. Processing foods with added chemicals further changes their nutritional profile. Hydrogenating the fat in peanut butter so that the paste and oil do not separate may be a boon to those who hate to stir the stuff, but the process creates a form of fat that definitely is not good for us.

Another factor is what goes into the food to begin with, such as from the soil that the food is grown in. Soil has (or does not have) nutrients. These nutrients are plants' food inputs. To some degree, whatever is in the soil goes into the plant. Fertile, nutrient-rich soil produces more nutritious carrots than depleted soil does. Soil aside, plants grown in the wild contain certain components or chemical compounds—**phytochemicals**—that can provide ingrown or natural protection from predators and disease. Artificially selected plant species often have had these protective phytochemicals bred out, not purposively but as a secondary effect because chemical pesticides allow plants without them to survive. Yet many phytochemicals found in wild species have antimicrobial or antioxidant properties (and provide more robust colors and flavors). These phytochemicals often are healthful for human beings to eat, but lacking in typical industrial supermarket produce (see Pollan 2006).

For animals, something similar happens, but in this case in addition to being linked simply to inputs (think of grass-fed versus feed-lot beef), outputs become important as well: how much the animal exercises affects the fat content of its meat. Wild game is much leaner than most supermarket meat, coming as it does from animals that not only romp and play more than feed-lot or factory-grown 'animal product sources' but that also run from predators. Indeed, the fat wild game does contain is proportionally more 'good' than 'bad' (saturated).

When we compare and contrast present-day human populations eating a modern diet to early humans eating as foragers would have, the difference to health is palpable. Biocultural diversity in rates of heart disease, diabetes, digestive disorders, and so forth reflect diversity in human subsistence patterns.

The Foraging Life

Foraging is not simply a dietary strategy; it is lifestyle. To better envision this way of life, consider the following example, which some have taken as representative, at least loosely, of the lifestyle that must have been experienced and enacted by our ancestors. It is drawn from data collected and descriptions proffered by Richard Lee (2000 [1968]).

!Kung Foragers

Lee worked with !Kung people to get data regarding work time allocation, nutrition, calories taken in, and so forth (note that the exclamation point printed prior to 'Kung' represents a clicking sound that we do not make in English). The !Kung live in the Kalahari Desert, which spans the middle-west of the southern-most part of Africa. The group of !Kung that Lee and colleagues worked with lived in the Dobe area, which had eight permanent water holes and, at the time of the research, fourteen independent and self-sufficient camp groups (see Figure 6.2).

The research was conducted quite awhile back, and things have changed drastically for the !Kung due to the incursion of others into

Figure 6.2
San people
in Deception
Valley, Botswana,
start a fire by
rubbing sticks
together, 2005.
*Photo by Isewell
via Wikimedia
Commons,
CC-BY-SA-2.5.*

!Kung territory and an onslaught of modernization efforts. At the time Richard Lee was able to stay with them, the !Kung were no strangers to the outside world. Yet they did subsist nearly totally via their foraging. As such, the old data are a vital source of forager lifeway information. In keeping with ethnographic tradition, we speak of the findings in the 'ethnographic present,' describing the snapshot they present as if current, so as to enliven the description.

Each !Kung camp fends for itself. Only rarely are foodstuffs traded between camps.

However, people move often, whether between camps or as camps themselves are moved to new locations in the region. Such shifts are a constant feature of !Kung life; most camps at least change locations with the seasons. In summer, when the rains come, there are many temporary water holes to camp around. In the dry season, camps cluster around the permanent water holes. Lee's fieldwork spanned nearly one and one-half years, which is important because of the impact of seasons on !Kung foraging life. Short studies, in which researchers buzz in and out of a community quickly, lack a comparable comprehensive perspective.

Diet

From any !Kung water hole, food is within walking distance, although the distance walked does increase to up to ten to fifteen miles in the dry season. In any case, people generally go to collect food every three or four days; a supply of no more than a few days' worth of food is usually on hand, and whatever has been hunted or gathered is always shared with the camp and so quickly dissipated. Even if people did want to store food, it would not necessarily last well. The !Kung have no refrigerators or pest-proof storage lockers.

Food kept on hand, however shortly, consists mostly of plant-derived items. The main staple item in the !Kung diet is the mongongo nut. A **dietary staple** is something that a group eats a lot of (variously bread, tortillas, or rice, for example). For the majority of human beings, now and throughout time, staple foods are not hunted: they are plant-based. Grown on trees, mongongo nuts are wrapped in a thin layer of edible fruit flesh. Lee reports that the nut is so abundant that more pounds of it rot on the ground than get eaten—even though mongongo nuts account for half of the vegetable diet of the !Kung, weight-wise. Average daily consumption is 300 nuts per person, and they are eaten throughout the year.

The !Kung also have access to eighty-four other species of edible plants, most of which are roots and bulbs or fruits, berries, and melons. Small mammals, birds, reptiles, and insects make up the remaining proportion of their diet.

No matter what else is eaten, mongongo nuts are central. On average, the nuts contain five times the calories and ten times the proteins of cooked cereal crops. The average daily portion yields about 56 grams of protein and 1,260 calories. When we include other food sources, the average !Kung intake is 2,140 calories and 93.1 grams of protein. By weight, meat accounts for 37 percent of this. Mongongo nuts account for 33 percent and other vegetable foods for 30 percent, putting the total vegetal contribution to the !Kung diet—pound for pound—at least at 63 percent; note, however, that plant food generally weighs less per cubic inch, therefore accounting for more dietary volume, than meat.

What does this all mean? Besides providing ample evidence for the argument that foragers subsist largely on plants, it also demonstrates that they are well-nourished, at least insofar as what has been measured. Today's United States Department of Agriculture (USDA) recommended daily allowance (RDA) of protein is 91 grams per 2000 calories; recommended caloric intake for an active adult woman of average build and nineteen to fifty years old is between 2,200 and 2,400; for an active man in the same category it is between 2,800 and 3,000. Note that for sedentary women it is 1,800–2,000, and for sedentary men it is 2,200–2,400. Both !Kung men and !Kung women are very physically active—but they also are generally of smaller stature than the average American. The RDA for them has been estimated at 1,975 calories and 60 grams of protein per person per day.

Lee does not distinguish in his 1968 report between men and women, and he does not say why not. We can infer that this had to do with his interest in health at the population (group) level, and the fact that both !Kung men and !Kung women led physically active lives. Moreover, the research Lee describes was actually carried out about half a century ago, when it might not have occurred to him to sort men from women when it came to the basics of nutrition. Although the older data are valuable because they precede mass globalization, which has affected the !Kung lifestyle, Lee's data-related decisions reflect the culture of science as it was in his day, for better or worse.

Hours Worked

Lee was interested not only in !Kung diet and nutrition status but also in the time and effort it takes to feed oneself and one's family on a forager diet, and here he does distinguish data by gender. He finds that while men and women both work a roughly equivalent number of hours per week, women (who focus more on gathering than hunting) bring home two to three times as much food as do men. Lee also found that the number of hours worked per week specifically gathering or hunting for food is what we would call 'part time.' In fact, it is twelve to nineteen hours—not even 'half time' by our standards. That is all it takes for the !Kung to get the food they need to eat. With the living so easy, children are not expected to contribute substantially to foraging efforts; this will not happen until they are married (between the ages of fifteen and twenty here). The very old and infirm, also, are well-provisioned by the able-bodied adults in the camps.

Lee's foraging figures do not include time spent making and fixing tools. Gross, based on Lee's 1979 report on same, put the figure for that at 6.3 hours per week, complementing his average of 17.1 hours per week in food-getting labor (1992, 265). Lee did not include housework (for example, food preparation, firewood collection), which Gross puts at 18.9 hours per week. Neither Lee nor Gross includes child-care.

Hourly workweek statistics for the U.S. and other workforces also ignore the time it takes at home to feed and wash, otherwise 'recreate,' or rejuvenate the workforce for another day on the job. Home-based maintenance of human resources is essential for production, but as it is not paid for directly by employers, it is left out of the waged labor ledger books. Moreover, and in any case, even with the labor necessary to recreate or maintain the means of production included in weekly work-hour totals, the !Kung, and those with comparable forager lifestyles, have quite a lot of time left over for leisure pursuits, including spending 'quality time' with extended family and friends (see Figure 6.3).

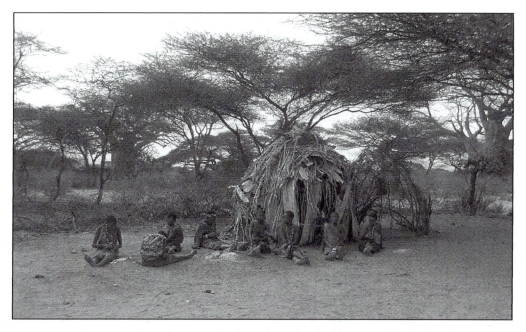

Figure 6.3 Hazda group at leisure in Tanzania. *Photograph by Rachel Naomi David. Used with permission.*

Food Options

Of course, at this point you might ask if the !Kung only work part-time getting food because there is not much food to get. Maybe they spend their off-hours lying around too weak to enjoy themselves. The answer to those contentions would be not at all.

It turns out that the !Kung, as we think was the case with our forager ancestors, are gathering only a small proportion of all the edible plants and animals known to them. Ninety percent of their plant-based diet is drawn from 23 Dobe-area species when there are 85 edible species known to be available. Animal wise, the local area has 223 species known to the !Kung, who classify 54 of these as edible yet regularly hunt only 17. Why? They are being picky, eating only what best appeals. They can afford to be picky because the environment amply provides. As per the nutrition data recounted earlier in this chapter, the !Kung get not only plenty to eat but good nutrition also—and they do it via the foodstuffs that taste best to them.

Seasonal Stress

I do not want to paint too rosy a picture here. Non-nutritional challenges that foragers face will be raised later; for now, however, I do want to bring into the equation Dobe seasonality. For most Western people, at least in the middle classes and above, there are no seasons. If we want strawberries in December, we can get them. Many of us have freezers, and some know how to can vegetables and preserve fruits. If something perishable (fish, fruit) is on sale or in good supply we can buy a bunch of it, freeze or otherwise preserve it, and eat it six or more months from now. We also have access, should we desire it, to non-local food, which extends quite substantially our menu of available eating.

If we lived as traditional foragers, our diets would shift with the seasons. Yet **seasonality** means more than this. Not only would what we eat change: how much we eat would also be affected. For many of the few foraging groups left in the world, the difficult season is much more difficult than it was for the !Kung when Lee stayed with them. This is largely due to the marginal lands they have been pushed into and incursions into their traditional foraging grounds. Be that as it may, even for some pristine foraging traditions there are certain times of the year in which certain kinds of food—and in dire cases all foods—become very scarce. Giving the body a cyclical break from constantly having to process certain kinds of nutrients can be a very good thing; but without food for long enough, people do suffer and can even die.

Exceptions to the Rule

Anthropologists are famous for taking exception. When people make generalizations about humankind, someone always manages an anthropological veto: "Not among the [insert group name here]!"

This does not mean generalizations have no value for enhancing understanding. It is helpful to know that some aspects of human life are universally found, more or less. However, we must be careful that generalizations remain a starting point for discussion and investigation rather than a stopping point. Otherwise, we are working with stereotypes, not generalizations (Galanti 2008, 7). Furthermore, the uncontextualized assertion that *all* foragers everywhere and throughout time have eaten a well-balanced, nutritious, and mostly vegetal diet (and without too much work having gone into it) would be plain wrong. In general, it has been the case—but not everywhere and not always.

Exploring an exception here is thus a necessary step in our learning. It will be illustrative not only of biocultural diversity but also of one of the key principles behind the initial evolution of such diversity: geographic variation in resources. People adapted to a myriad of environments when they spread out around the world, and although many ended up like the !Kung in a fairly hospitable area, others had to invent new ways of hunting and gathering better fitted to their new environments.

The Inuit Are Hunters

The Inuit, who live in the Arctic areas of lands now called Canada, Alaska, and Greenland, are one such group. As described by Ann McElroy and Patricia Townsend (2004, 15–28), the Inuit survive mainly by hunting. Besides the challenge of finding enough to eat in order to stay well-nourished in a region that supports almost no edible plants and very few animal species, the Inuit also must keep from freezing to death in a climate that is very, very cold.

Inuit bodies differ from !Kung bodies. As we know from Chapter 5, these differences are not 'racial.' They have to do with the unique geographic regions in which each group lives. The Inuit are quite stocky (although not fat), and they have somewhat shorter limbs. Their bodies have evolved to have less surface area than bodies adapted to hotter climates. They are, therefore, less subject to what in the Arctic habitat would be excessive cooling through too much body heat loss. There are internal differences as well, for instance in terms of the kind of fat Inuit have—they are able to keep making 'brown' or extra-body-warming fat into adulthood whereas other humans stop making it in childhood. Plus, Inuit blood vessels can very quickly dilate; quick dilation is helpful to reheat cold fingers and toes (McElroy and Townsend 2007, 81–82).

Bodily adaptations are helpful to be sure. Without a cultural boost, however, via adaptive clothing, shelter, and food-related practices, Inuit

Figure 6.4 Inuit in Canada killing salmon with spears. *Canadian Geological Survey, Frank and Frances Carpenter Collection, Library of Congress, LC-USZ62-112765.*

survival in the Arctic would be absolutely impossible. Yet survive they have done, for thousands of years, and very well indeed (until, of course, their invasion by industrialized nations). Although Inuit culture as a whole provides much to learn from—take for example the bone sunglasses or goggles that the Inuit make to protect against snow blindness, their social structure and childrearing practices, or their cosmological relationship with polar bears—here we will focus on how they make a living—how they subsist—in an environment where most of us, without Inuit knowledge and skills, would perish in a day or two (see Figure 6.4).

The Inuit do eat some plants. Berries grow abundantly in August, for instance. Inuit also gather and eat kelp or seaweed, sour grass, and sorrel. Other plants ingested by the Inuit are those dried for use as medicinal herbs and teas.

As a result of their minimal dietary dependence on plants, Inuit carbohydrate intake is relatively low: about one-fifth less than average American consumption, and maybe one-tenth of the average consumption of carbohydrates in poorer nations. Rather than subsisting mainly on plants, the Inuit subsist mainly on animals. They are hunters.

The energy costs of Inuit subsistence are large. About 160 days a year are spent in high-intensity activities such as hunting, trapping, fishing, and traveling by dogsled (which entails lots of running, pushing, and pulling). When in transit, a new snow house must be built or tent erected every night. Still, the Inuit spend more than half of their time relaxing, visiting each other, trading, having feasts, or building and repairing their equipment.

With such great reliance on animal sources for food, the Inuit protein intake is relatively high: about 200 grams per day. Fat, too, is eaten in what to the !Kung—and even to today's Americans—would be huge quantities.

Nevertheless, the Inuit are not fat or vastly unhealthy. For one thing, in part due to the work they do to survive, the Inuit overall are lean and muscular. Men expend between about 2,700 calories a day (3,100 in peak hunting periods). Women do not hunt, but they can burn nearly as many calories, particularly when carrying children on their backs, digging for clams, or breastfeeding. Remember that sedentary U.S. women and men—that is, most U.S. adults—only use 1,800–2,000 and 2,200–2,400 calories each day, respectively.

The nutritional content of the diet also contributes to Inuit health status, typified by healthfully low blood pressure, low cholesterol levels, and low rates of heart disease. How can that be with a diet so high in flesh foods? For one thing, the meat they eat is significantly lower in saturated ('bad') fats than present-day feed-lot beef, and higher in polyunsaturated fatty acid ('good'). The omega-3 polyunsaturated fatty acids found in fish and blubbery animals, such as seal, whale, and polar bear, add to the healthfulness of the mix. Moreover, while the foodstuffs eaten may be seen as low in vitamin C, when compared to the average American diet, the Inuit maximize access to the C that is there by eating meat raw; eating the plankton in the stomachs of raw fish, walrus, and caribou; and eating whale skin.

However, calcium intake is less than optimal in the Inuit diet. The ingestion of dried fish and bird bones as well as the soft parts of animal bones to some degree compensates for this (Inuit molars and jaws are relatively strong, which is adaptive when dealing with this kind of hard, crunchy food). Still, some Inuit do suffer from mild deficiencies in calcium, particularly in winter when sunlight is scarce, and so the body makes less vitamin D, for which exposure to sunlight is necessary. With less vitamin D, calcium absorption is less effective. In some populations a lack of sunlight without supplementation translates to rickets among children. Among the Inuit, who nurse their children extensively, this is not a problem (breast milk has calcium). Nursing women, of course, bear the burden of this, and Inuit adults as a whole—particularly those over 40—have an elevated risk of bone loss and fractures (high phosphorus intake contributes to this).

In summary, the Inuit subsistence strategy provides a reminder of the general lesson that there always are exceptions to the rule (in this case, the rule of a two-thirds vegetal diet). While the Inuit are mainly meat eaters, we also might have chosen a mainly vegetarian group to illustrate this fact—the general point remains the same. So, too, does the specific point made: what foraging constitutes depends on a group's circumstances. In this case, our focus is geographical circumstance, and what is therefore available. Internal biological adaptations, too, such as in regard to digestive processes, also come into play.

A follow-on thought regarding the 'ideal' diet is warranted here. There are people today who recommend adopting a dietary strategy that mimics the 'stone age' ideal. In general, this is sound advice; after all, many present-day substances passing as food or drink are unhealthful. Still, people interested in such a diet should proceed with caution. The apparent food equivalents might not actually be so equivalent when the methods by which they are grown or raised (for example, industrially farmed or wild) are taken into account. Moreover, the impact of diet is not singular; our activity levels and non-food as well as other factors mediate and moderate diet's effect on health.

Nourishing Babies

Our brief exploration of Inuit foraging considered some costs and benefits of nursing. This is actually a crucial topic. Most investigations or descriptions of subsistence ignore the fact that infants cannot digest, let alone chew, food like mongongo nuts or seal meat.

For the bulk of human history, babies have fed off the breast (see figure 6.5). The bosom's evolutionary success stems from the fact that it enhances survival of the population by providing nourishment for infants who otherwise would not (could not) eat. Breast milk has evolved in tandem with our need to give birth to such helpless offspring. We simply cannot gestate (grow and carry in the womb) babies until they are mature enough to eat solid food; giving birth to such big-headed creatures would kill us. Breast milk provides a way around this conundrum.

Perfect Nourishment

The nourishment provided in breast milk is perfect for infant needs, such as brain growth; plus, it is easy to digest, and this suits the infant's

Figure 6.5
Depiction of breastfeeding by Désiré François Laugée (1823–1896).

body, which is immature on all counts. Infants cannot digest the same kinds of food that even toddlers do.

Breast milk also is associated with leanness. When we compare children who were breastfed to those served formula, we find lower body mass and less propensity toward obesity among the breastfed group. Observational studies suggest that this is due to practices that surround breast and bottle feeding respectively. What seems to be happening is that bottle-feeding caregivers mix up the formula, fill up the bottle, and give it to the infant to suck from; so far so good. Now imagine the infant is full. Like the breastfed baby, he or she simply turns away from the (bottle's) nipple. However, if there is formula left in the bottle, the caregiver tends to try to get the infant to drink a little more. Formula is expensive, and many people prefer not to waste it. Formula-fed infants can thereby get overfed and used to feeding past the point of feeling full. This can have an impact that lasts into later years.

Perhaps because breasts are not see-through and the costs of making milk are not so transparent, women tend to leave breastfed infants alone once suckling concludes. Moreover, all things being equal, breastfeeding (nursing) women only tend to make as much milk as the infants in their care actually need. The body is set up to supply the volume of milk that the infant demands through a feedback loop that connects the infant's sucking (stimulation) to the breastfeeding woman's hormonal production system, and thereby her milk making propensity. The nurser and nursee comprise, in effect, one system. The more an infant suckles, the more milk a nursing woman makes. Breastfeeding women automatically offer a more or less perfect, individually tailored serving size for the infants in their care.

I have, until this point, used the phrase 'breastfeeding women' rather than mothers because not all women who breastfeed are genetic mothers to the infants they breastfeed. Some women who breastfeed are what we call 'wet nurses,' women who keep their breastfeeding capacity alive by suckling infants who are not their genetic progeny. Others may be biological relatives or even friends simply lending a hand (breast) with childcare. There is, in fact, an adaptive significance to caring for infants not directly your own, and that was discussed in Chapter 4. For now, let us focus on the biological benefits of breast milk and breastfeeding per se rather than worrying too much about which woman does it. At the same time, let us not lose track of the community-building function of sharing this task or honor.

In addition to nutrients, suckling infants also receive a certain amount of antibodies in breast milk, which help the infant to resist certain kinds of infection and disease. What happens is this: just as the infant's digestive system is immature, so, too, is its immune system. The nursing woman's immune system fills in for the infant's immune system while it is under construction. The **colostrum**—the yellowish fluid that infants can glean from the breast in their first day or so of life, before the real milk comes in (remember, milk production is stimulated through suckling)—is packed with antibodies.

Other Benefits

There are a few other benefits to breastfeeding that can be mentioned here. For instance, breast milk is sanitary. It has no time to be contaminated as it moves from the nipple to the infant's mouth. With formula, chances of contamination are rife. One must ask, "Is the water clean? Are the bottles clean? Are those nipples clean?"—questions mostly irrelevant to breastfeeding.

In addition to being good for baby, breastfeeding is good for mom. For one thing, breastfeeding seems to protect nursing women from certain forms of cancer. That includes breast cancer and possibly ovarian cancer. It also provides what one might term a 'natural' form of birth control because, while a woman is breastfeeding, she's often not ovulating. This has to do with the balance of hormones in her body. When ovulation is suppressed, fertility is suppressed. This can be beneficial to both the physical health of the mother and her mental well-being.

Notable here is the interconnection between the breastfeeding woman's body and the immature body of the infant. Parent and child form a system in which, as shown previously, much more than calories flow— and in which the flow is not just one-way. The infant's immaturity, related to the human need to give birth before the child is too big for passage, is offset by adult care. The parent's body can produce for the infant ample antibodies as well as foodstuffs. So, while the first three trimesters of our nine-month gestation are spent within the womb, a fourth is spent just outside of it. The infant freshly delivered still can be understood as if part of the mother's own body. Alternately, we might say that the boundaries between infant and mother are remarkably porous.

In any case, without breastfeeding, until very recently infants just would not have survived. If a population's infants did not survive, then that population would have perished. Breastfeeding was, by necessity, the norm.

Duration of Breastfeeding

Moreover, breastfeeding used to extend a lot longer then it generally does today. Data collected by John Whiting and Irvin Child for non-industrial societies (Whiting and Child 1953) showed that the average age for weaning (the age that a child is taken off breast milk completely) was two and one-half years old. By today's mainstream American standards, that may seem amazing. However in most cultures, for most of time, what we now call "extended" breastfeeding has been the normal way of life. Indeed, of the fifty-two societies in the study, only two practiced weaning prior to a child's first birthday.

In other words, the cross-cultural, pre-industrial norm was much more in keeping with current recommendations than most U.S. families are or can be today: the World Health Organization (WHO), for instance, recommends that we initiate breastfeeding within the very first hour of life and then breastfeed on the infant's demand, whenever the infant is hungry. We also should exclusively breastfeed for at least the first six months of life and continue to breastfeed until the child is at least two years old.

Why Do Patterns Vary?

Individual Reasons

If these are the recommendations, why is there variation? At the individual level, some variation is due to biology. Some women—and some babies—simply have a harder time with breastfeeding. For instance, an infant with a mouth condition such as a cleft lip or palate may have problems latching onto the nipple to suckle, or a woman may have a low lactation response, perhaps due to hormonal or nutritional factors. If a woman is having a hard time with breastfeeding, it is not necessarily a problem for the infant in a traditional culture because, as we learned earlier in this chapter and in the previous discussion of allocare (see Chapter 4), someone else who has recently had a baby may be more than happy to breastfeed the woman's infant. Today, of course, baby milk formula is available.

Cultural Variation

What about variation at the cultural level? In many cultures, breast milk is a valued, life giving substance, and breastfeeding is a valued practice. In other cultures, breast milk may not be seen as healthy. Some cultures teach that other substances are healthier for an infant. This is perhaps more likely to be true in modernizing, male dominated societies that equate formula with the modern West and disvalue almost anything to do with the female person or body. In certain situations, buying formula

Figure 6.6 Ruth Uchida, *left,* and Haruko Nagahiro, "milk kitchen girls" at the Heart Mountain Relocation Center, Wyoming, prepare formula diet for the center babies using sterilized bottles, 1943. *Photograph by Hikaru Iwasaki for the Department of the Interior. War Relocation Authority. U.S. National Archives and Records Administration. NARA record: 8464103.*

and the bottles needed to feed a baby on formula is considered a good way to show off one's wealth as well as to confirm one's commitment to modernity or 'development' (see Figure 6.6).

Other ideas that vary across cultures, and affect breastfeeding practice, include ideas about who has ownership over a woman's body: the woman or infant, or a husband or boyfriend? Such ideas play a very important role in determining who gets to manipulate whose breasts.

At the same time, some corporations are very interested in seeing our beliefs turn against breast milk. Some of these corporations have gone so far as to disseminate falsehoods about the value of formula. Some have been taken to court and been fined and told not to do this. Laws have been passed in order to protect women, infants, and families from this kind of exploitation.

And finally, for some people, breastfeeding simply is a luxury that cannot be indulged. When a mother works outside of the home and needs to keep her job, she may not be able to breastfeed exclusively for six months, let alone continue breastfeeding for over two years. This is despite the passage of laws, in countries like the United States, that support breastfeeding but are hard to enforce. Formula allows many women to stay employed or to gain employment as needed. Indeed, the initial introduction of formula was, in many areas, very important in terms of building both household and national economic stability.

Population Level Ramifications

In addition to the rise in numbers of women working outside of the home, the drop in breastfeeding that formula and the devaluation of breastfeeding supports also is correlated with rises in infant malnutrition-related illness and infant death rates. Death is sometimes due to lack of nourishment, as when poor people stretch formula servings by adding less powder and more water; it also can be due to mixing the formula with contaminated water or serving it via contaminated bottles or nipples. Moreover, we see less resistance to illness in infants with no access via breast milk to maternal antibodies.

Beyond child health issues, we also see increased population level rates of breast and possibly ovarian cancer in woman. Furthermore, we see populations themselves increase. In other words, there are more babies born. Why is that? With less breastfeeding, there is less ovulation suppression. Once ovulation restarts, conception becomes possible again and so birth rates go up when breastfeeding goes down. Bigger group size in itself has many knock-on effects, some of which we will explore in later chapters.

For now, a key point is that breastfeeding is an evolved, adaptive behavioral system—one that was selected for among early humans as part of a constellation of adaptations (including earlier birthing) because it enhanced our fitness. It also is one that, because of our capacity for culture, we now can contravene—for better and for worse. Another evolved adaptive behavioral system is co-sleeping.

Co-sleeping

What Is Co-sleeping and Why Do It?

Co-sleeping is simply sleeping in very close proximity—in the same room (where rooms are slept in), and sometimes in the same bed (or its equivalent, such as a hammock or whatever other furnishings a group uses for normal sleep). Co-sleeping does not include accidentally sleeping together, such as when a person sitting on a couch while holding a baby, say watching television, falls asleep with the infant. Co-sleeping is purposeful, and uses sleeping arrangements that are culturally recommended for legitimate, full-fledged sleep.

Moreover, co-sleeping is very beneficial for the interdependent parent-child system that we have been discussing. For one thing, co-sleeping supports breastfeeding. It's just a lot easier to breastfeed when junior is right close up to your breast than when, depending on your infant's cries to awaken, you have to get up, walk down a long hall, walk down the stairs, turn on the lights, fetch the crying baby, and so on.

Co-sleeping is very, very common across cultures even now, as it has been throughout history. In many societies, whole groups—not just nuclear families—sleep in the same hut, under the same roof, or around the same fire. Indeed, the whole idea of sleeping on your own, in your own room, or in your own bed, is very recent in historic terms. It still seems crazy in many cultures.

Why Separate Beds?

The idea of separate beds has a lot to do with an increase in the resources used for living; we now have bigger houses, more private space, and so on. It also has to do with culturally constructed understandings about spousal rights and obligations, such as are expressed through the idea of the 'marriage bed.' The mainstream U.S. idea that children should not be in this bed is culturally recent and relative. It ties into the cultural value placed upon independence, according to which we want children to grow up to be able to do things on their own, including getting themselves to sleep. We much prefer this to interdependence expressed by, for example, relying on another person's close proximity to help oneself relax. Mainstream Americans believe very strongly that, even for the littlest infant, sleeping in one's own bed fosters the kind of independent spirit that America was built on.

A review of current research suggests this may be wishful thinking. Our focus here, however, is on the basic adaptation of co-sleeping relative to breastfeeding and other health issues. One other health issue that it addresses is sleep regulation among infants, and this can be handled by fathers or other caregivers in addition to mothers.

A Developmental Bridge

Remember, infants are immature; their little bodies are still developing. When born, a baby cannot even hold up its head. It cannot grasp a rattle or hand, let alone a spoon or morsel of food (which of course it cannot bite or chew or digest anyhow). Infants certainly cannot walk or talk. Their immune systems are immature. So it should not come as a surprise that infants' cardio-pulmonary (heart and lung) systems, too, are as yet not fully developed. Even the brain's capacity to monitor how much oxygen is in the blood can still have some maturing to do once a baby has left the womb.

When the respiratory regulation system is working, a sleeper with low blood oxygen levels can arouse himself or herself to breathe more deeply. When the system is not yet ready to do that, death can result. Indeed, this seems to be a key cause of a horrible problem called 'Sudden Infant Death Syndrome' or SIDS. The baby simply stops breathing, cannot arouse itself, suffocates in effect, and is discovered dead.

As James McKenna and colleagues have shown (see, for example, 2005), the proximity of a co-sleeping individual to an infant serves, in effect, as a developmental bridge, giving the infant's cardio-pulmonary system time to mature. What happens is this: the sound and feel of an adult breathing and of his or her heartbeat help in turn to regulate an infant's breathing and heart rate. Studies in which people have had video cameras trained on them as they sleep demonstrate that when an infant is co-sleeping with its mother or another mature caretaker, each automatically assumes healthy and safe positions in which to sleep. Both are on their backs. Often the adult takes care of thermoregulation, for instance moving a blanket off an infant to effect a cool down (in addition to respiratory monitoring, thermoregulatory issues may be related to SIDS). With a breastfeeding woman, the infant can breastfeed at will. Neither adult nor infant must fully wake for breastfeeding or thermoregulation to happen.

Population-wide correlation studies have found that where co-sleeping rates are higher, SIDS rates are lower. We see this today in Japan, where sleeping with many people in the same room is quite common and culturally appropriate. In fact, Japan has one of the lowest SIDS rates in the world. We also see this kind of negative or inverse correlation (more co-sleeping, less SIDS) with overcrowding, such as in situations of poverty.

So although mainstream U.S. ideals suggest that co-sleeping is strange and abhorrent, it is the more typical sleeping arrangement around the world in different cultures and throughout time—co-sleeping is an evolutionarily beneficial or adaptive pattern. It is an adaptation whose benefits, like those of breastfeeding, mainstream Americans are canceling out with our cultural emphasis on independence.

Here and in places before, I have ended up talking about the present. Intrusion of the here and now demonstrates the contemporary relevance of knowledge about past biocultural patterns. What were those patterns? To recap, our forager ancestors breastfed extensively, and they co-slept. Moreover, they ate a diet consisting mainly (in most regions) of plant-derived food. No matter what regional constraints they were under, they developed ways of making a living and getting what they needed from the land and from each other.

Morbidity and Mortality among Foragers

It will be obvious by now that the availability of food and water is not the only thing affecting foragers' health. There are certain health threats entailed in extreme environments, of course, but there also are health threats entailed in the lifestyle that hunting and gathering entails more generally.

What are a few of these? Infection is a real problem, to start. Before the invention of pharmaceutical drugs like penicillin, a good proportion of people died from infections gone bad. A tooth infection, for instance, could end up taking a life. Dental problems in general have been a serious challenge for humans ever since we came into being.

Infection can also come on with an injury, as when gangrene sets in, and this can be lethal. There are also parasitic infections—infections with organisms like worms, and so forth. While parasites and infections are mostly only a problem now for people without access to health care, back in time they were everyone's problem. Chronic parasitic or bacterial infections, for example, led to many deaths and much disability and pain among foragers.

Early foragers also died (as foragers still do) from exposure to severe weather or predators, and from traumatic injuries. Today, if one falls out of a tree and breaks one's back or cracks one's skull, or is mauled by a lion, or experiences complications during childbirth, emergency medical services can be called upon. However, this option is a new cultural invention. There was no 911 service for early humans. Like infection, traumatic injury was a major cause of death.

As is true among the poor today, those foragers who survived often lived with the history of their infections or wounds written into their bodies: notable limps and missing fingers, teeth, eyes, and so forth differentiated individuals much more in the days before emergency—and plastic and reconstructive—surgery. On the bright side, most cultures accepted such diversity as part of the human experience and had no expectation of the kind of bodily homogeneity or 'perfection' projected in today's movies and magazines.

All told, however, the ethnographic record shows us that foragers are well-nourished for the most part, with a diet high in diversity as well as in plant-based foods. They also are fit; they have an active lifestyle in which the subsistence work of hunting and gathering are complemented with plenty of leisure time. Most early human foragers probably lived a similar lifestyle to that of the !Kung. We are reasoning backwards here: we do not have much beyond some skeletal remains and artifacts to show what life really was like for early humans. Still, it is reasonable to hypothesize that, like the Dobe !Kung, early human foragers probably did not have such a hard day to day life after all.

Dynamics of Human Biocultural Diversity: A Unified Approach by Elisa J. Sobo, 115–139.
© 2013 Left Coast Press, Inc. All rights reserved.

7

Agricultural Revolution: Another Great Divide

This chapter prepares you to:

- Characterize the subsistence continuum and explain the process of intensification it summarizes
- Explain why the Agricultural or Neolithic Revolution happened and how certain geographic regions supported this revolution better than others
- Describe the effects that the agricultural revolution had for sociopolitical and economic organization
- Describe agriculture-related health status changes seen in the archaeological record, including the skeletal record; describe also the biocultural effects of industrialized agriculture

Foraging is one of the most basic subsistence types that ever existed. The only thing more rudimentary is scavenging. Whereas foraging entails picking an apple off a tree, **scavenging** entails waiting until it drops to the ground. The forager hunts; the scavenger picks up carrion or road kill. Scavenging entails taking what nature has left out on the countertop, so to speak. With foraging, though, one's relationship to the environment is a little bit more manipulative. Compared to the scavenger, the forager has intensified. This chapter explores central consequences of intensification for human biocultural diversity.

Intensification

To begin to understand the concept of 'intensification,' have a look at the continuum presented in Figure 7.1. The various types of subsistence are represented on this continuum. 'Subsistence' refers to how a group makes a living—to how it gets food. **Intensification** is the process of doing more to get food: more intensified strategies manipulate or interfere with (construct) the environment more than less intensive strategies do. The more intensive the strategy, the more tools or technology will be used and the more complicated the technologies will be. More intensive subsistence strategies leave a bigger mark on the environment, changing it from its original form as food is extracted much more notably than less intensive strategies do.

Figure 7.1 Intensification continuum. Human manipulation of the environment is lower in subsistence forms to the left and higher in those to the right.

The continuum shown in Figure 7.1 organizes subsistence strategies from less intense to more intense, from left to right respectively. As we move to the right along the continuum, intensification increases. Beyond minimal-impact scavenging, then foraging, lies gardening or **horticulture** and herding or **pastoralism**. Sometimes there will be overlap; for instance, some foragers also may garden a little bit. In any case, horticulturalists and pastoralists are not settled farmers. Like foragers, they generally move around, following the seasons. Any gardening that is done is done without the help of irrigation, animal-powered plows, or fertilizer.

Once a people settle down and begin to manipulate the land with plows, special watering techniques, and fertilizers—once they embark on **agriculture**—a sea change occurs, and much of it stems from the very fact of settlement. This first occurred about 10,000 years ago. Of course, as with other key transition dates provided in this text, this one is more an *aide-mémoire* than a hard assertion; revolutions take time.

Regardless, the lifestyles that we see on the left side of the continuum pre-date that by a very long time.

Why Intensify?

Let's think of a foraging or pastoralist group that does a little casual gardening—a group that, more than likely, lives a life of fairly good nourishment and relative leisure. Why in the world would they seek a change? Why would they settle down? Three factors played into this development: (1) necessity, (2) geography, and (3) curiosity.

A Folk Model

From the mainstream American point of view, settling down supports a desire to acquire. Americans like having lots of stuff and often assume that others feel the same. We assume that foragers would not be nomadic if they had the option of owning material goods. We assume that agriculture opened up this option for such foragers (or pastoralists, or horticulturalists), who happily embraced it. The idea that people 'just want more' thereby comes into play in folk theories regarding the agricultural revolution—but not in scientific ones.

Agriculture is harder in many ways than foraging: intensification involves more labor. If we compare the workday of a forager to an agriculturalists' workday, we can easily see that the forager has the lighter workload. The workload goes up for peasant farmers and rises even higher for manual laborers who support industrial agriculture: while the peasant agricultural worker has been shown to work about nine hours a day, the industrial laborer works about eleven (see Schor 1993). Moreover, longer hours do not necessarily get people more of anything except exhaustion.

Climate Change Demands Adaptation

What really spurred the agricultural revolution in places where it happened? In a phrase, climate change. Societies became agriculturalists out of necessity. They had to. The climate changed, the **flora** (plants) and the **fauna** (animals) changed in response to that, and the typical diet constricted. People had to figure out another way to make a living in geographic regions that were affected.

Another argument beside the very strong climate change argument has to do with over-exploitation. Human beings simply over-exploited the flora and fauna where they lived. They pushed the ecosystem beyond its carrying capacity. This happens if, for example, people go fishing and take too many young fish: they deplete the fish population's reproductive potential. For that reason fishing crews today are advised to throw back little ones that when bigger can replenish the fish population. Over-exploitation probably played a part in supporting the agricultural revolution. However, most scientists favor climate change as the crucial driving force.

Indeed, climate data accord with archaeological data: approximately 10,000 years ago, the climate got warmer and drier in certain areas. People could not make a living through foraging anymore because the species on which they had depended could not survive well enough in the new conditions. People therefore had to figure out some other way to get food on the table. They either had to walk to another part of the world where the environment was more favorable to the foraging lifestyle or, if they were lucky, they found some candidate plants (or animals) for domestication right where they were. For some, simple supplementation via casual gardening—horticulture—or herding was enough. For others, full-out agriculture was the only answer.

Geography

In addition to necessity, geography had a lot to do with the agricultural revolution. Being in the right place—a location with great floral and faunal candidates for domestication—is something that Jared Diamond (1997) calls "geographic luck." The area now known as the Fertile Crescent, in Western Asia, near what is often termed the Middle East, had lots of such luck. Specifically, the Fertile Crescent covers Mesopotamia and the Levant. It is rimmed on the north by the Anatolian Highlands and on the south by the Syrian Desert (see Figure 7.2). Various continental plates come together in the Fertile Crescent in a way that supports a high level of biodiversity in the environment. Happily, what is true at the species level is true at the ecosystem level, too: biodiversity has a protective effect in the face of new environmental pressures (climate change included).

The Fertile Crescent was the first place in the world, at least according to current archaeology, that saw domestication of plants and animals. The population there when the climate changed was fairly dense, relatively speaking. Remember, human beings first evolved in Africa, and they first got to the rest of the world on foot via the Middle East. By the time the climate changed in the region, humans had in fact spread around the world, with populations growing sparser the further away they were from the African hub. Nevertheless, some of the other places humans lived seem to have experienced the great impact of climate change a little bit later. Moreover, they generally did not have the advantage of such relatively easy access to eligible domesticates (exactly what an eligible domesticate is will be discussed soon). For such reasons, the Fertile Crescent stands out.

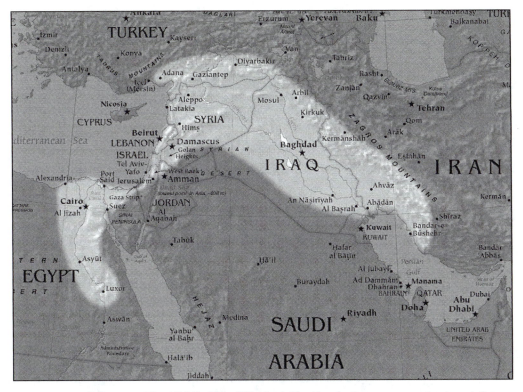

Figure 7.2 The Middle East (CIA Map). The Fertile Crescent runs in an arch from the Mediterranean Sea to the Persian Gulf; it is rimmed on the north by Turkey and on the south by the Syrian Desert.

Curiosity

Beside necessity and geography we must consider a third factor as having led up to the agricultural revolution. It is the case that climate change led some people to need to figure out another survival strategy. It is the case that certain people were geographically lucky or got geographically lucky when they walked away from where they had been and found a new place to live. It is also the case that curiosity comes into play because people, being people, are curious. Take, for example, my dear childhood friend, who got a toy car wheel stuck up her nose when very young. You likely know someone who did something similar with a pebble or a bean. Why do children stuff things in their noses? They are curious. We are curious: we are human beings. We can in fact learn from what we observe after applying our simple curiosity. The child who gets the rock (wheel, bean) stuck up his or her nose once very rarely does so again.

We use our senses to get to know the world. Our capacity for empirical observation—for looking around (feeling, tasting, smelling, hearing, and so forth) and noticing things—has been key to our species' survival. Fortunately for us, we noticed at a certain point that when a seed falls on the ground, it sprouts. Then we noticed that we could control where a seed was planted to sprout. We noticed that if we put it in one kind of soil it grew faster than if we put it elsewhere. We noticed that if we chose the sweetest fruit to eat, and seeds from that fruit got planted, the bushes or trees that grew up from them too yielded sweet fruits. Why did we notice? Because we are curious.

So curiosity was a very important part of the equation that supported the agricultural revolution. A group might need to intensify, and a group might find itself in a spot where there were lots of eligible domesticates; but if that group—that population—did not have innate curiosity and the intellectual capacity to derive meaning from what was observed, agriculture could never have happened.

Those Who Could, Foraged, Until They Could No More

Of course, not everybody had to adopt an agricultural lifestyle; not every population was pushed to do so by climate change. Many groups maintained a foraging subsistence strategy or lifestyle. It may be that some developed cultural adaptations to control population size and therefore to reduce pressure on the land's then dwindling carrying capacity. For example, they may have realized that there were more effective forms of family planning than lactation- or breastfeeding-induced ovulation suppression alone.

Further, in some areas, the need for agriculture did not come into play until more recently. In fact, up until about 2,000 years ago, about half of the world's population foraged for a living. Today the proportion doing so is nearly nil.

Why did the numbers of foragers keep going down if 10,000 years ago those people who needed to intensify did so? The answer has to do with the power of settled agriculturalists to expand and their evolved desire to do so. In some cases, foragers have basically been pushed out by farmers who expanded into the foragers' territory to take over fertile lands. Nowadays, of course, there are corporate interests at work—logging companies, mining companies, and so on—pushing foragers out of business, either literally or through the pollution they produce, which poisons the foragers' livelihood.

Eligible Domesticates

What was it about places such as the Fertile Crescent that enabled the adaptive emergence of agriculture? It was proximity to eligible domesticates:

people who lived there were close to plants or animals that were actually able to be domesticated: plants and animals with certain features.

What were those features? Jared Diamond (1997) provides a good review. Beyond being visually noticeable, and tasting and smelling nice, he says, the best plants to domesticate are those that can be stored for a long time without rotting. They likewise have seeds that can lie dormant until planted (seeds that do not sprout when in storage)—and that can be easily gathered in the first place.

Wild plant species generally have seeds that are easily dispersed. Wild wheat seeds actually shatter from their shafts in the breeze when ripe. A person walking through a wild wheat field can, by simply brushing a plant's stalk, cause shattering to happen. Wild wheat seeds (also called grains) are therefore hard to gather—except when the wheat grows in a mutated form, wherein seeds are more firmly attached to the shaft.

Early farmers would have inadvertently selected for this shatter-proof variety, despite its natural selection-related disadvantage. Why? Because it would have been easier for people to gather grains to eat from such plants. They also would have selected for softer, plumper grains or seeds

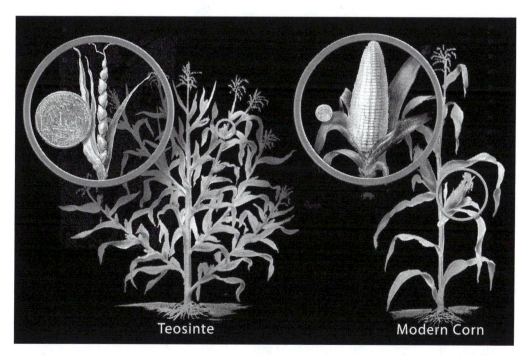

Figure 7.3   Ancient and modern corn. Among other things, ancient corn has multiple branching stalks and more, smaller ears with fewer, harder kernels; modern corn has a single, non-branching stalk and more, softer kernels on fewer, larger ears. *Courtesy Nicolle Rager Fuller, National Science Foundation.*

rather than the tough-coated seeds that do best for the plant in the wild (see Figure 7.3). To be further viable as domesticates, crops should be fast-growing and have a good yield. Like wheat, other seed grain crops like rice, barley, peas, maize (corn), and millet fit the bill. For better or for worse, however, these originally were found only in certain geographic regions.

With animals, Diamond notes, similar rules of thumb apply. The animals that reproduce and grow to maturity quickly are more favored than those that like, say the elephant, have only a few offspring that mature slowly. Animals good for domestication also must be people-friendly (they should not, for instance, eat them), and they should be sociable with other animals and docile, too. Animals like pigs, goats, sheep, and cattle are thereby excellent candidates; lions, zebras, and so forth are not. Eligible animals were only available, originally, in certain geographic regions, adding to the geographic luck of those living there (Diamond 1997).

Revolution at What Cost to Human Health?

Agriculture was good. It allowed some populations that might have perished to survive. It eventually helped fuel the scientific and industrial revolutions. However, it was not without costs. The agricultural revolution had ramifications for the health and well-being of populations who fostered its practice.

Agriculturalists live a lifestyle that differs vastly from that of foragers. To practice agriculture, a people must settle. They can no longer live as nomads. They must stay in one place.

Exposures

An **exposure**, technically speaking, entails close proximity to a chemical, **pathogen** (germ), radioactivity, or extremes of weather. One of the costs of settling down to do agriculture is an increase in a group's exposure to environmental toxins. This might, for instance, include lead, which can lurk in water, soil, and food as well as in dishes and utensils. Lead, taken in, can poison various organs and tissues; it interferes with a number of bodily processes. One key symptom of lead poisoning is abdominal distress; others include headaches, irritability, and seizures. Coma and death also are possible. Lead poisoning is especially dangerous to children because it interferes with the growth of the nervous system; it can cause developmental problems, if not coma or death.

Indeed, archaeologists have identified sites where lead poisoning is thought to have been common due to its existence in the environment.

For example, in locations where clay used to make pottery had high lead levels, people would have suffered.

If foragers had passed through such an area, they might have been exposed to a little bit of lead, but not much, because the foraging lifestyle typically is nomadic. Foragers follow the seasons, moving to different parts of the environment when different animals are going to be there or different fruits come ripe. They move around. They might come back and revisit camps on a seasonal basis, but they do not stay put—and this minimizes chances for toxic-level exposures.

Agriculturalists on the other hand do not move. They stay in the same place for a lifetime. The longer people stay in one place, the more likely it is that local environmental toxins are going to penetrate their bodies, and the more likely it is that they will suffer ill effects.

Speaking of lead, a more modern case bears telling of here. It takes us to Devon, England. Devon is famous for, among other things, apples. Readers who have been to Devon know that cider—apple juice, hard (fermented) or soft—is the drink of tradition.

Turning apples into cider is one way to store them or to extend their shelf life. Agriculturalists all had to figure out modes of storage. Storage containers can, it is true, contaminate food; this easily could have happened if, for instance, the cider jugs were made from clay that contained lead or from pottery whose glaze was made using lead (lead makes a glaze shiny). However, as Ian Maxted explains, the case of lead poisoning's history in Devon, and its link to the apples, was not so simple as that (2000).

Lead poisoning once was so common in Devon that it was referred to as the Devon or Devonshire Colic. Colic is a kind of stomachache often attributed to babies that cry a lot (and seen much more frequently in modern Western cultures than in others where, for instance, babies are always held and breastfed on demand). Devon Colic was a disease of the adult population mostly. The colic is first described in writing in 1655. The people in Devon at that time saw a connection between drinking cider and the colic; they generally explained the connection as due to cider's natural acidity.

Common sense long held, but in the 1760s one Dr. George Baker hypothesized that the lead in the beverage was to blame for the colic. The abdominal distress of the colic also was seen in known cases of lead poisoning. Sure enough, it turned out that lead was used in making cider: it was used in the presses and in the cleaning process. Chemical tests done by Baker confirmed high levels of the contaminant.

The published results were not kindly taken to by cider makers. Nevertheless, eventually the cider press makers did remove the lead and

Devon Colic became a disease of the past. Important here is the point that doing something more 'intense' with the apples than just eating them off of the trees had dire health costs. Of further interest is the fact that Baker's data were rejected by those with a profit motive. This is something that tends to happen in capitalist economies, themselves made possible only after the rise of agriculture—but we shall return to that later.

Effects on Skeletal Health and Growth

Beyond using material artifacts to get a sense of how agriculture affected health, skeletal remains and teeth are also revealing. For instance—and for much of this and the next few sections we will rely on the expertise of Clark Spencer Larsen (2006) and Mark Nathan Cohen (1989)—the hard work of agriculture is written into the skeletal joints. Agricultural populations demonstrate higher rates of degenerative joint disease, reflective of more demanding activity patterns. They also have thicker bones—again a reflection of a more physically demanding life (exercise stimulates bone manufacture).

When we look at the skeletons of ancient agricultural populations, as well as more recent ones such as those of Devonshire in the middle ages, we see shorter stature than is common amongst foraging peoples. Short stature can be a sign of increased stress on the body, often pertaining to childhood malnutrition (see Chapter 3).

We also see something called **porotic hyperostosis**: bones that are more porous than they should be. This is a sign of iron deficiency, which triggers the expansion of bone tissues that form red blood cells. Scientists are still in discussion as to what causes the deficiency. It could be nutritional stress, but there is a competing hypothesis regarding immune defenses. It turns out that certain pathogens thrive on iron in the blood, and so some people are hypothesizing that the body withdraws its iron in order to create an environment in which those pathogens are unlikely to survive.

Whatever its origins, porotic hyperostosis in a group of skeletons signals that bodies were under stress, and more porotic hyperostosis is seen in agricultural populations worldwide. So while there are benefits to the settled farming life, skeletal evidence shows that there also are costs. What we start to see cropping up in agricultural populations is an assemblage of problems representing what we might call the diseases of settled life.

Some of these diseases are written into the teeth. We see declines in dental health in settled agricultural populations. We see it in terms of simple decay due to changes in the diet. Gruels or porridges made of grains and rice and so forth stick to the teeth in a way that forager staples do not, causing more dental caries (cavities). "Virtually everywhere that human populations made the shift to agriculture," writes Larsen, "saw a rise in the frequency of carious teeth" (2006, 93). We may wonder at Larsen's use of the term "virtually"; he himself notes, based on his literature review, that "rice does not appear to be as cariogenic as other domesticated plants, especially maize." In other words, folks who eat more rice should not have as many cavities as folks who eat American corn. Culture does make a difference. Having said that, Larsen goes on to note that there are some indications that even populations subsisting mostly on rice may have had higher dental caries rates than foragers, and in any case their teeth suffered in other, agriculture-related ways.

For instance, we can see general nutritional stress, common to agricultural populations, written on the teeth they have left behind. When people are under nutritional stress the enamel does not get laid down the same way as it does with a healthy well fed individual. For instance, we see **enamel hypoplasia**, which means that there is less enamel in some places on the teeth than there would be on healthy, evenly-nourished teeth. When a person is not getting ample nourishment, the body may not have enough raw material to contribute to building teeth, in which case we see pits and lines on the teeth as signs of the body's deficit. We see enamel hypoplasia indexing physically stressful periods. It also can index a struggle with infectious disease which, as the next chapter shows, can plague settled populations. We certainly see more evidence of tuberculosis, syphilis, leprosy—infectious diseases that affect the skeleton—in agricultural populations than amongst foragers. But for now, let's keep the focus on nutrition.

Causes of Nutritional Stress

People who have a settled agricultural lifestyle have poorer nutrition than foragers (Cohen 1989; Larsen 2006; see also Pollan 2006). This links back to (1) a lack of variety, (2) nutrient loss in the foods eaten due to absorption as well as storage issues, and (3) lower nutrient levels in the foods to start with due to some domestication outcomes and soil problems.

For one thing, when compared to foragers, settled agricultural populations generally have a much narrower diet. They have come to depend on the products of their agricultural labor for subsistence. If the weather does not support a good crop, or if stored crops rot, agriculturalists are subject to famine. Moreover, the lack of dietary variety even in good times narrows the range of nutrients that their bodies are taking in. Remember, foragers eat lots and lots of different foods. We have a saying that variety is the spice of life; it is not just that: variety is vitally necessary for a well-balanced diet that provides all of the nutrients human beings need.

A lack of variety leads to malnutrition among agriculturalists. Nutrient loss does, too: it turns out that many of the major cereals contain within them certain phytochemicals that hinder the absorption of other nutrients. Diets heavy on corn (maize) can foster a niacin deficiency because certain phytochemicals in the corn inhibit niacin's absorption. Wheat eaters often are zinc deficient. A diet highly dependent on rice can lead to vitamin A and thiamine deficiency.

On top of problems that people might have because of phytochemicals inhibiting absorption, they also can suffer from storage-related nutrient loss. The longer we store food, the more time that food's nutrients have to break down—meaning that the same food becomes less nutritious as it ages.

A third factor affecting the nutritional status of agricultural people is the simple lack of wild foods in the diet. Wild foods generally are more nutritious than foods that we have raised domestically. Wild plants tend to have more of their own internal phytochemical protections against predators, for example. Domesticated agricultural plants do not need these as much because humans work to protect them, using pesticides and other tools. It turns out, however, that many phytochemicals that protect plants against predators also have benefits for the human body. In addition to giving a fruit or vegetable more intense flavor, and deeper color, some phytochemicals have antimicrobial effects in the human body just as they do in the plant. They can therefore offer us some protection against illness or infection.

Finally, agricultural crops are lower in nutrients because agriculturalists use the same lands again and again and again. When land is not allowed to lie fallow or helped to regenerate, the soil can become very depleted. Wheat or rice or millet—or carrots or apples or celery and so on—that is grown in depleted soil simply is not as nutritious as a crop grown in rich fertile soil. Plants feed on nutrients in the soil, and if they're taking in fewer nutrients from the soil because the soil has fewer nutrients to offer, humans will in turn take in fewer nutrients when they eat those plants.

There is a 'wild' food chain as well, of course. A wild plant takes in nutrients from the soil in which it took root. Rich, wild soil fosters more nutritious plants. Animals living in the wild eat these wild plants (and sometimes animals), thereby taking in a wider variety of nutrients than do domesticated animals. This is one reason that wild meat is more nutritious than, say, beef from cattle raised in a feedlot eating hybridized domesticated corn.

Moreover, meat that comes from wild animals is very lean in comparison to the meat of domesticates. Animals out in the wild on their own get more exercise than domesticates, running as they must to keep away from predators, or running to catch their own prey. Domestic animals don't need or get to do so much running or moving, and so they are fatter than wild ones. In addition to being fatter, domestic animals also have a higher proportion of what nutritionists would call bad fat—the kind that our doctors warn us against eating too much of. This is partly why today we have much higher rates of cancers, heart disease, and so forth than we used to.

Even our dental health is in some ways worse now than in the past. It is the case that we have modern fluoridated toothpaste and so forth, but we also have a much, much softer diet than even our agriculturalist ancestors did. On top of that, we eat vastly more sugar. The sticky, sugary foods that we eat exacerbate the human tendency to get cavities.

A Negative Correlation

All in all, there is a negative correlation between intensification and good health. That is, as a given population moves from a foraging lifestyle to a settled agricultural lifestyle, its health declines: as intensification goes up health goes down.

Happiness also might decline, or at least certain kinds of unhappiness are more evident, perhaps due to increased competition for resources. In some regions, injuries linked to interpersonal violence, as seen in marks left on skeletons, increase over time as agriculturally productive lands grow scarce. Organized warfare also seems to increase.

Forests are cut back to clear new land, erosion occurs, and some species die off due to changes in the ecosystem. Other species, now in increased or closer contact with human populations, share out their parasites and other pathogens. Larsen in fact goes so far as to refer to the agricultural revolution, itself driven by environmental changes that served as evolutionary pressures, as an "environmental catastrophe."

While settled agriculture is in many ways a wonderful human invention, it can and often does have a downside in terms of human health (Cohen 1989; Larsen 2006). We may scoff on first reading Larsen's dramatic description of the agricultural revolution as catastrophic, but the long-term, knock-on effects of this change, which we now are recognizing, actually have been just that.

Industrialized Agriculture

Settled agriculture involves a more intensified relationship to the environment than foraging. With agriculture, humans are doing a lot more to manipulate the environment. When settled farmers began to engage in 'cash cropping' or 'monocropping'—that is, growing just one crop, and lots of it, to sell as a commodity for cash rather than to eat—and when they began to use highly mechanized means to do that—**industrialized agriculture** was born.

Streamlining Natural Systems

The industrial approach to farming requires us to break more of the eco-logical links that nature had set up environmentally. There are all kinds of links in natural systems. On the old-fashioned farm, chickens eat worms busy fertilizing the fields. Then, the chickens defecate in the fields, further fertilizing. Cows go out and eat the grass now growing, and they in turn defecate. Flies fly onto these cow patties and lay eggs; chickens eat them when they hatch to maggots, and so on. This is just one of the many, many circles of flowing energy and matter that we find in the complex adaptive system the old-fashioned farm represents.

The industrialized food production system strives to be more linear. It takes animals, like pigs and cows, and plants, like corn and wheat, out of the complex systems in which they have evolved in order to streamline their 'production' (see Figure 7.4). In doing so, it creates new challenges, and we cannot always predict what these emerging challenges will be. Take, for instance, the fact that feedlot cattle cannot roam and therefore end up standing, sometimes knee deep, in their own waste. Such cattle can succumb to diseases carried in the manure or resulting from stresses that feedlot life engenders. One answer to this is to flood their feed with antibiotics. This has knock-on effects, such as contributing to antibiotic resistant germs that make humans who eat or handle the cattle very sick (for more on resistance, see Chapters 8 and 9).

The point here is that how a population subsists or gets its food (cul-ture) can have a very real impact on that population's bodies (biology), leading to human variation. Some of that impact comes directly via diet; some comes from labor patterns, evidenced in certain kinds of wear and tear on the body, for example. Another portion of the impact comes indirectly: it comes through the way a subsistence type affects the envi-ronment when longstanding systems are altered.

Energy and Soil Use

For example, the energy costs of certain kinds of subsistence go beyond the human calories required to find or produce the food. This is seen in

Figure 7.4 Chickens in an industrial coop on a factory farm. *Photo by ITamarK. Public domain via Wikimedia Commons.*

the vast amount of fuel it takes to bring food to the table in a modern industrial food production system. We add fertilizer, which today is made largely out of fossil fuel. We also use gasoline and diesel fuel to transport food, and we use energy to process and package it.

Another way to think about energy costs is to look at loss instead of input. Everything that we eat initially draws its energy from the sun. The sun shines down on the earth, and the corn grows. The corn grows, it is harvested, and it is made into feed for cattle. Cattle eat the corn, and eventually the cattle are slaughtered to become hamburgers. In that process, energy is actually being lost. By the time we go from sunshine to corn to cow to meat, we have lost 90 percent of the energy that the sun originally had contributed (Pollan 2006).

Beyond fuel energy costs lie the cost of soil degradation and erosion. When we plant a crop on a piece of land, year after year that land gets leached of its nutrients. Land also loses soil. Our pastures and our range lands are losing soil year by year due to overgrazing and overgrowing.

In sum, intensification helps populations to survive; but it also has costs. These include not only energy costs, such as of fossil fuel and human labor, but also costs borne directly by the environment and variously by human populations.

New Food Movements

Before moving on, it is worth noting that food and social activists are gaining traction in efforts to turn things around, in part by relinking ecological chains that industrial approaches have denied and doing so in a way that takes the lessons of large-scale settled agricultural life as well as of industrialized agriculture into account. Take, for instance, the bio-dynamic, slow food, and farm-to-table movements. Each approach has its own emphasis, but all promote the local production of a wide variety of regionally appropriate foods, in season, and with sustainability in mind. All foster multiple and complex relationships between soil, plants, and animals—humans included.

Biodynamic farming in particular emphasizes our mindful stewardship of the earth and preservation of biodiversity. Industrially derived fertilizers and laboratory-produced seeds are shunned in favor of polycultural methods in which farms and gardens exist as their own potentially self-sustaining complex adaptive systems (Biodynamic Farming and Gardening Association 2012).

In general, new food movement efforts favor small-scale enterprise that allows for face-to-face relations. Most keep issues of equity in mind, too, working to ensure that healthful foods are equally available to all people. These new food movements seek to change not only the way we subsist, and thereby our health, but also the way we experience and care for our world.

Intensification and Social Change

The Four Ss

As we have just seen, the agricultural revolution changed much more than simply our diet. With agriculture we settled down or adopted **sedentism**. As we became sedentary we also built granaries or silos (see Figure 7.5). We had **surplus** to store—extra food to save for use well after the harvest. The existence of surplus supported **specialization** and a related division of labor among certain segments of the population in areas other than food production, such as the healing arts, or in the crafts of pottery or architecture. A growing but segmenting population also set the stage for **social stratification** or hierarchy. In later chapters we will see what all these things meant for biocultural diversity per se. However first we need a deeper understanding of what these terms mean and a better grasp of their implications for the human experience.

Figure 7.5 "Perfect Granary," an exceptionally well preserved Ancestral Puebloan (Pueblo III) structure from 1100–1300 AD. Natural Bridges Archeological District, Natural Bridges National Monument, Utah. *National Park Service record #42SA6788.*

To bring us to that point, we have to talk about something that happens to people when they settle down. They have more babies. Why? For one thing, people do not need to worry anymore about having too many children to comfortably travel because they are no longer walking over long distances (remember, our feet used to provide our only carriage). With many young children, that kind of travel can be very, very hard (one can only carry so many infants and young children at a time). Nomads tend to have very small families. People who are settled, however, have the ability to build larger families. They also have the need: more children means more help on the farm.

When families get bigger, social groups or societies get bigger, too. When we look at the ethnographic and historical records, we can see certain patterns in the way that people organize themselves, both politically and socio-culturally, as their populations grow. One of the ways that they do this is through new political arrangements.

Among foragers, people live in small family-based **bands**. Bands may come together also, into **tribes**, in which a number of bands have ties that knit them together. For instance, bands can draw spouses from one another. This often happens when pastoralism or horticulture begins to enter the foraging mix and a few fixed villages, often seasonal, emerge. Sometimes tribes are knit together when representatives from each otherwise generally nomadic band form a cross-linking group, perhaps based on gender or age. This may occur, for instance, in regard to religious or spiritual needs relating to the yearly cycle. In any case, and aside from gender or age distinctions, bands and tribes (whether foragers, pastoralists, horticulturalists, or a mix) are essentially egalitarian groups with very informal governments or decision-making processes; conflict resolution is done between people, who must simply work things out. Peer or cultural pressure is used to enforce decisions and encourage pro-social behavior. Leadership, if any, is achievement based (see Table 7.1).

When people intensify to the point of practicing settled agriculture, political arrangements also change. Government, decision-making, and conflict resolution get more and more formalized. In the political form

Table 7.1 Four Forms of Political Organization and Their Implications on Selected Aspects of Life

	Band	Tribe	Chiefdom	State
Government, decision making, conflict resolution	Egalitarian, Informal	Egalitarian or Achieved leadership, Informal	Hereditary, bureaucracy, power to enforce	Centralized, multi-part bureaucracy, legal system
Settlement pattern	Nomadic	Semi-nomads, Fixed village	Fixed villages	Fixed villages and cities
Division of labor, Exchange system	No, Reciprocal (gifts)	Not really, Reciprocal (gifts)	Some, Redistributive (tribute)	Lots, Redistributive (taxes)
Add-ins			Public architecture, luxury goods, slavery, literacy	

called a **chiefdom**, hereditary bureaucracy, in which people inherit their positions, emerges. Bureaucrats and people who have the power to enforce decisions now exist. Indeed, chiefdoms have chiefs. In **states**, which are even larger, generally, than chiefdoms, bureaucracy is centralized, with multiple parts, and a formal legal system exists. While bands and tribes are highly egalitarian, chiefdoms and states are not—just as they have multi-layered political systems, they have multi-layered social structures. There are people at the bottom of the social structure and people at the top. In other words, there is social stratification; there is a hierarchy of classes. Often, chiefdoms or states have a slave class. There also will be public architecture and luxury goods. These all depend to a large degree on the existence of a surplus, which can be used to support specialists— people who need not labor in the fields but who can (or who are forced to) devote time to non-subsistence related pursuits. Among the elite, who patronize the artists, we also often find high rates of literacy in chiefdom and state societies (again, see Table 7.1).

Social Evolution and the Organic Analogy

What this brief discussion of political organization encapsulates for us is a pattern of social evolution. That pattern was described for us more than one hundred years ago by Herbert Spencer. If that name sounds familiar, it is because Herbert Spencer is the person who really coined the phrase "survival of the fittest," and he coined that phrase in relation to economics. Spencer wrote for the *Economist*, which today is still a major weekly news magazine. As a journalist and prolific scholar, Herbert Spencer spent a lot of time thinking about the changes that were going on all around him in England (see, for example, 2002 [1874–1896]).

What he saw was the impact of the industrial revolution. He saw multitudes moving from the countryside into the urban areas to work in the factories, where many people fell by the wayside. The factories were filthy and dangerous; they had a horrible effect on people's health. People suffered from poverty and poor living conditions. There was a sense among the upper classes of uneasiness or anxiety about what was going to happen as a result. The question of how social order would be maintained was very near and dear to the heart of social theorists of the day. Many scholars looked, ethnologically or comparatively, to other societies for answers.

Using a form of the comparative method, Spencer concluded that societies, like organisms, go from a state of being homogeneous to a state of being heterogeneous; that is, they go from a state in which everybody is fairly equal and similarly educated to a state in which people have distinct and different roles and hold disparate class positions. We know now, of course, that without evolutionary pressure, a band can stay a band forever. However, given a need to intensify, a society will evolve from simple to complex in terms of its social structure; it will go from being homogeneous to heterogeneous. Now, rather than being a **simple society**, with interchangeable parts, it is a **complex society**. This is the core of Spencer's theory of social evolution, which is all summed up in what he calls the "Organic Analogy."

According to the organismic or **organic analogy**, societies are like organisms. Although we now know that evolution is multilinear, in the late 1800s, when Spencer wrote, the ladder analogy held much sway. Given the climate in which Spencer was doing his thinking, his theory holds that a society will become, over time, more complex (structurally speaking). It will evolve parts with specialized functions; it will evolve, for example, a formal legal system and a formal educational system. All these parts emerge over time as that society evolves.

Compare the evolved society to the human organism. As in a complex society, there are many different parts in the human body. Our bodies are very complex; all of the parts have very specialized functions. Some body parts lead (for example, the brain). Other parts follow (for example, doing as the brain says). Complex bodies—and societies—must stay organized or they will not survive. In contrast, in single celled creatures, the inside parts are quite homogeneous. In this, the simple organism is like a band society.

Social Cohesion

Even as Spencer wrote, another social theorist, Emile Durkheim, also was thinking about the question of **social cohesion**, or of how societies stick together once they grow big (as they do with an agricultural subsistence base). Durkheim focused on the challenge of what has been termed **social condensation**—which is when larger groups break into smaller factions. In social condensation, a group of people condense or pull together to make their own little group, much as water condenses into droplets on a cold drink bottle (see Figure 7.6). This can happen in bands and tribes; some families simply pull up stakes and move on if the group gets too large or conflicts arise. It does not happen, however, in chiefdoms and states. People wondered why.

Durkheim's answer focuses on the **division of labor** or, more specifically, extra-familial economic specialization (1997 [1893]). I make this distinction because every family—even in a band—has some kind of

Figure 7.6 Water vapor condensing or coming together to form droplets on a bottle, much as people in one society come together to form subgroups. *Wikimedia Commons, GFDL User:Acdx.*

internal division of labor, if only one based on age and gender. However, in a band, or even a tribe, a family or household unit is self-sufficient. Each household unit is, in effect, interchangeable or equal. The same is not so in chiefdoms or states, where extra-familial or household-to-household labor divisions exist. Durkheim's division of labor insures cohesion as opposed to chaos or condensation.

How does it do that? It does that because the division of labor ensures household-to-household interdependence. In other words, single households or groups cannot cleave off to form or start a new version of the society from which they came because they are not independent units. In a foraging group, most skills are equally distributed between households; foraging groups are homogenous in that regard. However, the existence of a surplus among agriculturalists helps to ensure that some people will specialize. Some households will not have to focus on subsistence per se. One could specialize in pottery, another in healing, another in making cloth or making buildings. The model for supporting these specialist households could be communal, or they may trade goods and services for food. Alternately, they may be slaves in the service of an elite (who, again, lives off the surplus).

The division of labor, as Durkheim calls this, means everybody is responsible for a different bit of the labor continuum. Responsibility for goods and services is distributed throughout the group differentially. If José knows how to make toothpaste, and Jenny wants to brush her teeth, she is very dependent upon José. Likewise, if Jenny knows how fix cars, and José needs his car fixed, he is dependent on Jenny for that service. With a division of labor we have interdependence, and that trumps or diffuses any threat of social condensation. Because of the division of labor, people depend upon one another in particular ways; from this, social cohesion emerges. Thus, larger societies with a division of labor are in effect self-ordering.

The relationships between people in large, complex societies are strong because without them we would struggle if not perish. They also often are single-use or single strand, one-dimensional relationships; indeed, they are not really 'relationships' but rather instrumental exchanges organized around practical needs. They are not person based but follow the office; that is, the toothpaste maker, or auto mechanic, or teacher is not important because of who he or she is as a person but what he or she does (his or her 'office'); therefore, one toothpaste maker is, in theory, as good in regard to supplying your toothpaste as is the next. There is no long-term commitment signaled in making a toothpaste purchase from one toothpaste maker. The use of money for such an exchange can give even more concreteness to its one-time, single-use value. However, such exchanges do create a strong and cohesive social fabric (see Table 7.2).

Table 7.2 Forces for Social Cohesion

Simple societies (Bands, Tribes)	Complex societies (Chiefdoms, States)
Multiplex ties	Single-strand ties
Expressive relations	Instrumental relations
Enduring bonds	One-time exchanges
Reciprocity	Market exchange; Redistribution

In contrast, in a simple society (a band or a tribe), using the term 'relationship' to describe an exchange makes sense. This is because it is generally the case that in a simple society the people engaged in the exchange also are kin. They must see each other often and in different roles as well; in other words, their ties are multiplex. Their exchanges have a temporal dimension. That is, they last or endure over time, often through a cycle of exchange actions in which tallies or records may not even be kept mentally because what is being shared is thought of as if a gift. This type of exchange relationship, infused as it is with emotional or expressive value, and meant to be of long-standing duration, entails what theorists call a **reciprocal relationship**. Each positive act is met in turn with another, in continual back and forth movement. Importantly, return gifts given need not be material. They can include help with housework or childcare or a hunting or house-building effort. Each gift given contains a little bit of the giver and cements the ties between those doing the exchange and often between their families or households as well.

With **market exchange**, however, which is the main type of person-to-person exchange that characterizes chiefdoms and states, links between what is given and the giver are broken. The use of money signifies this directly, although it also can be true with barter (trading). Chiefdoms and states also use **redistribution**, in which money or goods are taken from one subgroup for use with another, or for use society-wide. Redistribution would be, for example, when the tax collector collects dollars from individuals or households and spends them on things meant for the public good, like education or to build roads. It also can entail hoarding by a self-serving ruler. In any case, the link tax dollars once had with the taxed individual or household is severed with redistribution.

Importantly, the redistribution of funds or surplus grains or goods in chiefdoms and states often is used to support hierarchy, or social stratification. This means that some groups have more rights and entitlements than others. Some people are members of the elite, and some people are on the bottom of the social heap. Often, food production falls on the backs of those people lowest on the social scale.

Sociopolitical Diversity and Biocultural Diversity

What are the implications of this chapter's discussion for us? Diversity in sociopolitical arrangements supports biocultural diversity of various types, from diversity in the kinds of physical stressors that write themselves onto the body, to diversity in the types of diseases to which people might be exposed or vulnerable. We will be exploring such considerations in the next few chapters. I shall therefore close here with a quick review of the social processes that we have been talking about.

When people settle down to do agriculture, their populations increase. They grow surplus food, and more people plus a stored surplus allows for specialization. When people start doing different kinds of labor, with some of it considered of higher value and some of it considered of lesser value, the door is open for social stratification, and stratified social systems emerge. Knowledge of these processes helps us to better understand the biocultural diversity that came into being with the rise of agriculture and that persists in the world around us.

8

Epidemics: Human Settlement and Sociality

This chapter prepares you to:

- Correctly use basic epidemiology terms in discussing biocultural diversity
- Characterize the human immune system's role in the body-environment relationship
- Use the disease ecology perspective to explain how the agricultural lifestyle contributed to the spread of infectious diseases
- Characterize ways in which local ecologies contribute to population diversity in vulnerability to certain diseases

We human beings have an evolved ability to protect ourselves from the intrusion of unwanted visitors (viruses, bacteria) through our immune systems. Barring individual dysfunction, the basic mechanics of the immune system are the same for everyone around the globe. However, because specific hazards to our bodies vary from place to place, groups of people have over time developed different functional variations in immunity through the process of natural selection. As a result, there is also variation, from population to population, in our vulnerability to certain diseases.

Much of this variation has to do with where a given group ended up geographically and how that group managed to feed itself. Related settlement patterns and subsequent changes in social organization (culture) affected human health (biology) and led to diversity between and even within populations. This chapter explores the biocultural interrelationship implicated. To do that properly, we first have to understand the

165

immune system and some of the basic principles of **epidemiology**, the study of disease distribution and its determinants. With firm foundations there, we will go on to examine the relationship between settlement, related patterns of sociality, and the spread—or not—of epidemic disease in various populations.

Remember, our immune systems have evolved, and diseases have evolved right along with them. Some diseases did not even exist in particular places prior to the human occupation of these regions: they exist *because* of us settling there. Adding insult to injury, humans don't just cause new diseases, we also cause more diseases. Epidemics as we know them never could catch hold until our populations grew large enough to support them. Why do large populations support epidemics? The answer has a lot to do with the biocultural changes that came with agriculture. The ecology of epidemic disease is, in many ways, a human creation.

The Immune System

Keeping Out and Dismantling Invaders

The immune system is made up of organs, tissues, and cells that protect our bodies against invaders. Beginning with the skin, organisms and materials that are not meant to enter the body are mostly kept out. At a deeper level, our white blood cells, or **leukocytes**, work to keep intruders at bay. These cells circulate in a system or network of lymphatic vessels as well as in blood vessels, and they are stored in lymphoid tissues and organs, such as in lymph nodes and the spleen. Lymph nodes are found throughout the body, but many are concentrated in the armpits, neck, and groin; the spleen sits above the stomach under the ribs on the left side of the body. The spleen filters the lymph and generates antibodies and lymphocytes, which are one kind of leukocyte.

There are two kinds of leukocytes: **phagocytes**, cells that eat or chew up invaders, and **lymphocytes**, cells that help our bodies remember and recognize previous invaders and support our bodies in destroying them. This is how they work: invading cells are covered in unique substances called **antigens**. A key subset of antigens works somewhat like enemy flags. Sensors in the immune system detect these flags and raise the alert (invader, invader!). Another part of the immune system responds by creating **antibodies**. These are special proteins that can lock, selectively, onto the invaders' antigens and effectively disarm them. They do this by either neutralizing the invader or calling other cells, including sometimes phagocytes, to eat or dissolve the invaders. This only works, of course, if the invaders can be recognized.

EPIDEMICS: HUMAN SETTLEMENT AND SOCIALITY | **167**

Immunization

Vaccination or inoculation shots—immunizations—work by giving our bodies the knowledge that enables them to recognize certain diseases quickly and fight them off immediately if they do invade. Most immunizations do this by introducing a dead or attenuated version of an otherwise dangerous invader into the body. The immunized body, now experienced with that disease, is primed to recognize and react to the real thing and is thereby protected.

Immunization of this kind was first tried in the West in the later 1700s. By that time, some had observed that milkmaids exposed to cowpox seemed immune to smallpox. Smallpox, named after the small boils or skin pustules it often causes, can be deadly; cowpox generally cannot, at least to humans. However, because the germ for smallpox is quite similar, from the human immune system's perspective, to the germ for cowpox, anyone previously exposed to cowpox could fight smallpox off more effectively than could people not previously exposed. More milkmaids survived smallpox epidemics for this reason. Edward Jenner, an English surgeon who had taken note of this, carried out the first vaccination upon his gardener's son, using pus from a milkmaid's cowpox eruptions. This history is embedded in the word vaccination: *vacca* is Latin for cow.

Technically speaking, there are two forms of immunization: vaccinations use look-alike germs; inoculations use real ones. A smallpox vaccination would have used cowpox while a smallpox inoculation would have used actual smallpox germs. In fact, long before Jenner invented the vaccination, people in China, India, and Turkey used inoculations to stem smallpox's tide. They would either grind up smallpox scabs and blow the powder up a person's nose or cut a person's skin (sometimes into a vein) and jab in fluid from the pustules of someone whose smallpox case had been mild.

Europeans who had heard of this form of immunization and were willing to immunize at all in fact preferred it over vaccinations. This is because people did not like the idea of mingling cow and human fluids. They would happily eat cattle and even suck beef bones of marrow, but the direct mingling of cattle and human fluids through the vaccination process struck much of the European population as bestial, so for quite some time they would have none of it (see Figure 8.1). Similar beliefs cause some people to reject vaccination and inoculation today. Another source of rejection is misguided faith in an infamous—and fraudulent—study, now discredited, that linked autism to childhood immunization practices (the author was stripped of his medical license once the fraud was exposed).

Figure 8.1 Cartoon inspired by the controversy over smallpox innoculation; drawn by James Gillray; published in 1802 by H. Humphrey. *Library of Congress, Prints & Photographs Division, LC-USZC4-3147.*

Three Kinds of Immunity

Immunity brought on through vaccination or inoculation is called **adaptive immunity**. There are two other kinds of immunity: **passive immunity**, which infants acquire from breastfeeding because antibodies are passed along via breast milk, and **innate immunity**, which is inherited. Innate immunity is passed to offspring in the process of natural selection, so it is tempting—but technically incorrect—to call it adaptive. Adaptive immunity instead involves the same kind of adaptation we talked about when we discussed developmental adjustment in Chapter 3. If we remember this little inconsistency, we won't confuse the adaptive and innate immunity.

Innate immunity is a bit more complicated than adaptive immunity, too. Whereas adaptive immunity happens to specific individuals through the act of inoculation or vaccination, innate immunity is a process that occurs over time in a population. For example, if a population is ravaged by a smallpox epidemic, a proportion of people tend to survive. Although some survive because they simply aren't exposed to the disease (the shepherd, for instance, who is off herding sheep on a mountaintop when the epidemic sweeps through the village), others survive because they are resistant to the disease for some genetic reason. As a result of the epidemic, then, the percentage of the population that is resistant to the disease is greater than it was prior to the epidemic. Likewise, resistant individuals are contributing a greater proportion of genes or alleles to the gene pool of the next generation because non-resistant people died and cannot do so. We can therefore expect the next generation as a whole to be a bit less susceptible to smallpox, and sure enough, history shows that smaller proportions of the population die from smallpox in generations following epidemic exposure to it. Nevertheless, in the case of smallpox at least, innate immunity is not a fast or comprehensive process: subsequent generations are by no means completely protected. Humans have not yet come close to evolving complete immunity to it, so smallpox has remained a terrible threat.

Immune System Deficiencies

In addition to diseases that overcome otherwise healthy immune systems, there are other problems, called auto-immune disorders, that can compromise our immunity. For instance, an immune system that lacks certain kinds of lymphocytes, or lacks enough of them, cannot effectively fight disease. Certain medications can weaken the immune response in this way, as can germs such as HIV, the human immunodeficiency virus. By contrast, sometimes an immune response can be too strong, even turning its weapons on the body itself. This can happen, for example, when the body over-reacts to common environmental materials and develops rashes, swellings, or other symptoms well known to anyone who has suffered from allergies.

Sometimes, allergies are the result of the training—or lack thereof—that the immune system receives when young. A study conducted by Erika Von Mutius in 1999 helped fuel a major shift in our understanding of the immune system. Von Mutius compared allergy and asthma rates in what used to be East and West Germany in 1999, hypothesizing that those in the East, which was historically poorer and dirtier, would have higher rates of these ailments. To her surprise, she found exactly the opposite to be true.

Children living in polluted areas of East Germany actually had lower rates of asthma and allergic reactions than did those in 'clean,' modern West Germany. Why? East Germans used more daycare; their children were around numerous other children. They also were around more animals.

Based on these differences, Von Mutius developed the **hygiene hypothesis**, which holds that the immune systems of children exposed to more microbes have greater tolerance for the irritants that trigger asthma and allergies. In other words, hyper-sanitation leaves the immune system with so little to do that it is primed to jump on the first bit of flotsam it encounters, no matter how harmless it might really be (Public Broadcasting System 2001).

Infection

Even the healthiest immune system cannot always fight off intrusions. One kind of invader that causes trouble is the communicable or infectious disease germ. Simply put, an **infectious disease** is a disease resulting from the presence and—this is key—the activity of a pathogenic microbial agent that can be communicated from one person to another.

Agents, Hosts, Environments

Germs—bacterial, parasitic, viral, or otherwise—act on the human body to make us sick; they generate pathology in the body. Because germs actively do this, epidemiologists call them **agents**. Agents need **hosts**: they need bodies on which to live; they need bodies to infect. Out there alone in the world, agents cannot survive. On or in a host, they can—but agents make host organisms sick.

How is a disease or germ transmitted or communicated to a host? Often this happens directly from host to host. Humans are great germ communicators because we are so social. We constantly reach out to shake hands, give or receive hugs and kisses, and touch each other socially. In some sense, our very sociality is our downfall when it comes to infectious disease.

For their part, the germs help ensure their own transmission through the symptoms they provoke in us, such as sneezing. In the case of tuberculosis, for instance, a cough and the droplets that come out of the nose and mouth facilitate the spread of germs. In diseases that lead humans to form a rash, the fluid in pustules, which should eventually burst, is the vehicle of transmission. The germs in that fluid are gambling on the chance that someone else will come in contact with them. If the student sitting next to you has a rash on his arm and he unintentionally brushes against you, or if someone who just sneezed into a hand turns (with that hand) a doorknob that you are about to touch, the agent will have hitched a free ride to the body of its next potential host.

This approach that examines the germ's point of view and experience of disease is a disease ecology perspective. Disease ecology focuses on the germ, the environment in which it lives, and the way in which it makes the most of its particular niche or habitat.

The agent and the host, however, are not always directly connected. Sometimes, the link between them is formed by a **vector**. Examples of common vectors are ticks, fleas, and mosquitoes. Agents also may get to hosts via an **environmental reservoir**—something in the environment that holds the germ, for instance soil or water. While a vector is bringing the germ directly to the host, an environmental reservoir is holding the germ until the host happens to come in contact with it. Epidemiologists also sometimes speak of **reservoir hosts**: these living organisms are both hosts—infected themselves by the agent—and reservoirs that can communicate their disease to other living beings. A diseased rat, for example, may bite and infect that pesky child next door.

Figure 8.2 The epidemiological triangle demonstrates the multidirectional relationships between host, agent, and environment. *Illustration by author.*

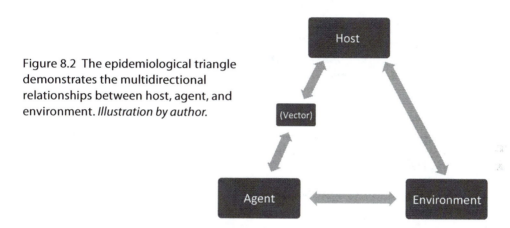

The infection process can be seen as based in the relationship between a host, an agent, and the environment (which is, of course, quite humanly created). This trio often is termed the **epidemiological triangle** (see Figure 8.2).

Epidemiological Vocabulary

Infectious disease is not the only kind of disease out there. Non-infectious diseases exist, too; for example, cancer and diabetes. A non-infectious disease is not caused by an agent or germ. We should also distinguish between chronic and acute diseases. A disease that comes on very fast and then (if it doesn't kill its host) goes away fast is an **acute** disease. The common cold and 24-hour stomach flu are good examples. By contrast, a **chronic** disease, such as HIV/AIDS, lasts for a long, long time. One way to remember the difference between these two types of disease is to note that chronic stems from the name of the Greek god of time *Chronos* (think of the word 'chronological').

These are terms we need to know to foster our exploration of the relationship between biology and culture as expressed through health. Another term that we need to know is **endemic**. An endemic disease is one that is native to a given population or area. It has been there for a long time and exists in balanced coexistence with the host population. 'Demic' comes from *demos*, also used in 'demography,' and even 'democracy'; 'demos' means 'the people.' The prefix 'en-' means inside, so an endemic disease is one that's inside the population—one that the people in that habitat have simply learned to live with. Malaria, for example, is endemic in certain areas.

On the other hand, an **epidemic** disease is initially not inside a community or people but rather comes down upon them—and spreads very quickly and very efficiently among them ('epi-' means upon or over, as the epidermis or skin lies over our bodies). In epidemics the agent usually moves from host to host—jumping among people, spreading itself very quickly, and infusing the population with the disease that it brings. Polio, for example, swept over the United States just after World War II, sickening huge numbers of people and leaving many paralyzed and many dead.

Figure 8.3   Original map made by John Snow in 1854 for On the Mode of Communication of Cholera. Cholera cases are highlighted in black. *Rewardiv at en.wikipedia, public domain via Wikimedia Commons.*

A **pandemic** is basically an epidemic that has a global toehold; it is an epidemic that's not just sweeping over one village, local region, or nation, but reaching all around the world. As we learned from the spread of the H1N1 (so-called swine flu) virus late in the first decade of this century, something that is pandemic or even epidemic does not necessarily kill the majority of the population. Technically speaking, these terms refer not to severity but rather to how far and fast a disease spreads. However, traditionally when people use the terms epidemic and pandemic, they do refer to acute diseases with high mortality rates.

The science of epidemiology emerged in response to some of the epidemics that were spreading through Europe, specifically England and France, during the 1800s. Just about any introductory epidemiology text will cover the London cholera epidemic, recalling the way Dr. John Snow cleverly traced it back to a water pump on Broad Street in London. How did he do that? He identified and mapped out the households in which people were getting sick (see Figure 8.3), figured out where their water came from, and found a positive correlation between getting water from the Broad Street pump and contracting cholera. It took some persistence, but finally he did convince public health officials to close the pump. The number of cases fell almost immediately.

Settlement and Sickness

This story brings us back to our discussion of how settlement patterns and subsequent changes in social organization (culture) lead to diverse human health outcomes all around the world between and even within populations (biology). Recall that bacterial and parasitic infections were a big problem for foragers. They continued to be a problem for settled agriculturalists. Other health problems persisted everywhere: traumatic injuries such as those that can be sustained during abnormal childbirth, from long falls, or from fighting during warfare or in the context of domestic violence. On top of all that, settled people began to suffer, and suffer greatly, from epidemic diseases— bubonic plague, tuberculosis, smallpox, cholera, measles. Why were epidemic diseases such a problem once people settled down? Three key reasons were malnourishment, more reservoirs and vectors, and mercantile trade.

As we learned in Chapter 7, many settled agriculturalists were not that well-nourished. Sometimes whole groups were malnourished, but often groups within groups, particularly the lower classes, were more malnourished than the rest. When a person is malnourished, the immune system cannot respond as quickly as it should; a malnourished person is not as well protected as he or she would be otherwise.

Figure 8.4 Professional rat catchers in 1900 in Sydney, Australia, display their work. *Photo by John Degotardi for the Department of Public Works, Australia. CC-BY-2.0.*

Also, when people are sedentary, there simply are more reservoirs. There are more people to act as reservoirs in settled groups, which fosters host-to-host communication, such as in the case of tuberculosis, or TB, which is borne in the droplets emitted when an infected person coughs. In addition, settlement is followed—at least in successful agricultural groups—by stored surplus food. Food stored in silos or granaries attracts vermin such as the notorious rat. Stored food is associated with bubonic plague because of the way that it supports rat populations.

Rats themselves are not actually the bubonic plague vectors for human infections; that is the job of the fleas that the rats carry. The rats are only a reservoir host. In fact, while controlling the rat population does serve a purpose (see Figure 8.4), some theorists argue that it was not until the host rat population was decimated by bubonic plague that people became susceptible. At that point, fleas jumped off of dying rats in search of better (that is, living) hosts and found humans. So if dead rats begin filling an area, the humans who live there should be quite worried; whatever killed the rats might be brought by the dead rats' ex-fleas to new human homes.

Bubonic plague, or black plague, is known as such because infected people would often develop buboes—big lumps like giant boils. These big, often blackening lumps generally form first in the groin, in armpits, or on the neck. These locations are home to lots of lymph nodes where white blood cells—infection fighters—cluster. The buboes turn black

Figure 8.5 Plague scene ("Epidemics Die Pest"). *Image from the History of Medicine collection, National Library of Medicine, 101405382.*

when they have gone septic or necrotic—when they are full of decay. If a bubo bursts, the pus that escapes will communicate the infection to others; whole towns sometimes succumbed (see Figure 8.5). In the past doctors commonly opened buboes, believing this would let out the bad juices (here is culture at work) but thereby spreading plague germs (and biology was affected).

However, the real culprit in spreading the plague was trade. The spread of the black plague in the Middle Ages tracks exactly to known shipping and caravan routes. Rats carrying fleas with plague got on ships; ships went from port to port; and rats went from town to town, bringing with them their fleas. Since then we have made trade much, much faster, beginning with train and then truck routes over continents and then with airline routes that link continents together by mere hours. Germs are jet-setting, too.

Disease Ecology

A Germ-Centered Approach

When we examine infection from the perspective of the agent we can understand how infection, as horrifying and grievous as it may be to people, is essential for the survival of each agent species. Disease ecology offers a germ-centered approach that we can use to situate biocultural diversity in a systems context.

Disease ecology focuses on the immediate or proximal environment, including a germ's immediate habitat, and on the context in which such germs are spread. Disease ecology is a narrow framework in the sense that it takes a very close-up, on the ground or in the weeds look at disease transmission. Were we to move from talk about trade routes to explore how different societies are linked together—and how geo-political relations, power structures, political economy and so forth fostered a disease's spread—we would be taking a helicopter perspective—one too broad to be termed 'disease ecology.' Think of a camera operator on a movie set. The camera would be observing events at a wide angle, drawing back to take everything in. Conversely, a disease ecology perspective calls for the close up; the camera lens narrows the focus right down to the germ, the disease, and the mechanics of how infectious disease spreads. This is the angle we take when we talk about how the immune system may be weak in a malnourished individual, allowing for infection. We also take the disease ecology perspective when we talk about the number of reservoirs in a farming settlement or focus on livestock, which we shall now do.

Zoonosis

When people settle down and farm, they don't just raise barley or wheat or millet or rice or corn. Farmers also can raise cattle, goats, sheep, pigs, and any other animals they domesticate. Many domesticated animals have been associated with the spread of epidemic disease among human populations. What happens is labeled zoonosis. A **zoonotic** disease is one that originated in an animal species. Zoonosis always was possible, but when human beings started to raise livestock, we broadened the possibility that something that infects animals would jump the species barrier and infect us. It also appears that zoonosis happened even to plant farmers; clearing land, for instance, may have supported the emergence of varieties of malaria that make people sick because it brought humans into closer contact with the chimpanzees who carried another strain of the disease (see Cormier 2011; Rich, Leendertz et al. 2009).

Normally, a germ that can infect a particular species cannot necessarily infect another because each species is a little bit different. In this, speciation is a bit like sexual reproduction in being a source of variation (see Chapter 2), although the latter serves only the single species by ensuring diversity in the gene pool while the former serves life in general by ensuring diversity at the level of the genus and beyond. The passkey a germ needs to enter and infect one species can be very different than the passkey to enter and infect others; some species have inbuilt resistances. For example, certain germs are well-fitted to thrive within (infect) pigs, but because of physiological differences between pigs and human beings, often those germs cannot infect us.

This status quo of germs and hosts holds only until the day that a germ mutates in just the right way to gain a toehold in the human body. This is what seems to have happened with malaria long ago and, more recently, with the H1N1 virus. Through this simple evolutionary process, a new humanly infectious disease is born. The longer a human group has handled livestock in farming or dealt with a particular kind of farm vermin, and the closer that physical contact has been—the more bodily contact those animals have with the farmers—the higher the likelihood that one day a zoonotic mutation will happen.

Population Size

Zoonosis is necessary, but not sufficient, for a new disease to become epidemic. Epidemics require a large population to support the transmission of a disease from generation to generation. There must be survivors. In a small population, if everybody is wiped out, so, too, is the particular pathogen that killed them, as there are no more hosts to keep it alive (unless, of course, it can jump the species barrier again). In a large population, however—particularly in one that reproduces sexually and thereby exhibits variation in the gene pool—chances are that some people will be immune to the new germ. They will go on to have children and provide the germ with a whole new host population.

Why do some people survive and how much protection will survivors pass to their children? Some people survive because they have innate immunity—a variation in their genes that confers upon them a resistance to the disease. If they live, their children are likely to carry forward that particular form of genetic resistance and be innately immune to the disease. You may be wondering how these children can provide a new population for the germ to infect. There are two answers. First, not all of them will be born immune because very few traits are 100 percent inheritable. Second, other forms of immunity (adaptive, passive) may have led to their parents' survival.

Smallpox

This entire scenario played out again and again so that, over time, European populations intimate with cattle were, in aggregate, less vulnerable to smallpox than were populations that had never been exposed to it. A traveler from a cattle-raising community might bring the smallpox to a town without cattle and inadvertently wipe it out entirely. Likewise, a traveler—a merchant, conquistador, colonialist, missionary, or any other—might carry smallpox to another nation or continent. The result could be absolute decimation.

Native Americans, of course, had never been exposed to either smallpox or cowpox prior to the invasion of the continent by European parties. Following European colonization, Native Americans, lacking any form of immunity, innate or adaptive, to the smallpox germ, died in droves, despite various precautions (see Figure 8.6). In the Northeast this period has been called The Great Dying.

Some scholars have wondered why the opposite did not take place—why the invaders did not sicken and die from indigenous American diseases. As Jared Diamond has noted (1997), although the Europeans had cattle, people living in the Americas had herd animals as well; in South America, for instance, they had llamas. There were huge differences, however, in the ways that these populations related to their livestock. For example, domestic llama herds were smaller than European cattle herds. Further, people in South America did not milk their llamas, nor did they live very near to them, so they never established that kind of physically close contact that the milkmaids and other Europeans had with their animals. In Europe, cattle often provided a source of body heat for the household; cattle might live in the same quarters on a house's ground floor. Put simply, Europeans were just closer to their animals—to their skin, to their feces, to their milk. The South Americans and their animals were not so close. They did not have the opportunity

Figure 8.6 Burning of a Navajo hogan occupied previously by a smallpox victim, near Indian Wells, Leupp Indian Reservation, Arizona, c. 1890–1990. *National Archives and Records Administration.*

to foster zoonotic diseases equal to smallpox; they did not coevolve with a disease to which they would be a bit immune but the Europeans would not. This example of biocultural diversity had huge implications for global history (Diamond 1997).

Recap and the Red Queen Redux

With settled life, and concurrently larger populations, disease is more likely to spread. No longer dispersed, agricultural people live in close quarters and very nearby to large numbers of other people. They tend to live near their own waste as well. Agricultural systems result in and are sustained by food surplus, which attracts vermin and therefore supports more disease. Living with livestock has the same result. Polluted water supplies compound health challenges. Foragers, by contrast, generally roam about the countryside, leaving waste behind. In permanent settlements, however, health problems pile up.

To better comprehend the knock-on effects of settlement, let's extend the Red Queen metaphor. As we saw in Chapter 2, the Red Queen keeps moving because the world around her is moving all the time. Human culture and human biology work much in the same way, adapting to new environmental pressures or challenges. Importantly, many of the challenges humans face are in fact humanly created. Changes in social organization and subsistence patterns foster the spread of disease. In changing the way in which they made a living, our ancestors made much more intensive and extensive changes to the environment than they ever had as foragers. In turn, they had to change—change their cultures (for instance inventing such things as housing that would last more than a season and synchronizing their belief systems to their new diets) and their bodies (for example, to digest lactose or gluten, or to resist new diseases). Our ancestors had to adapt to the effects of the changes they brought about; so do we, and so do other species. The world around us is in constant motion.

Cholera

Cholera is a horrible intestinal illness brought about by a bacterial pathogen that became very well adapted to thrive within large, settled human populations. Cholera epidemics have swept through the Old World numerous times in the past. By the 1800s, cholera had penetrated the New World, too, as the reach of Old World powers grew. Typically, 50 to 70 percent of those stricken in an initial epidemic of classic cholera will die (Joralemon 2006, 33).

Cholera's main initial symptom is the sudden onset of intense unabated diarrhea. This brings about immense dehydration: the body is excreting so much it cannot replenish its fluids, which is why it dies. Diarrheal excrement, of course, is helpful for the cholera germ because, if uncontained (and containing it can be quite difficult due to the nature of the illness), it serves as an excellent way of moving that germ between one host and the next. Water or raw foods can provide a reservoir; if the germ gets into another host before it expires, the cycle will continue. Ceviche, a raw shrimp and fish dish, is a very good conduit for cholera transmission if the ingredients come from water infected with the cholera pathogen.

The bacteria that cause cholera do so by excreting a substance that, to human beings, is very toxic. The substance disrupts the regulation of water and salt across the wall of the intestine. Under normal circumstances, nutrients (including salt) and fluids move across the lining of the intestine. One of the biological mechanisms that regulate the flow of nutrients and water across the membrane is a particular kind of **transmembrane conductance regulator**, or TR. TRs move (conduct, transfer) things across the intestinal membrane, regulating the flow (see Figure 8.7). When this regulation process is thrown off kilter by toxins secreted by the cholera-causing bacteria, the affected individual will be sick, not only because fluids and nutrients are insufficiently moved out of the intestine and into the body, but also because

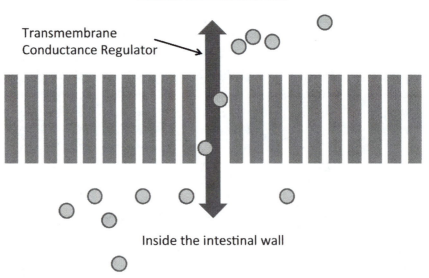

Figure 8.7 Transmembrane conductance regulator (TR); TRs regulate the flow of targeted materials across cell membranes. *Illustration by author.*

fluids are pulled out of the body from outlying tissue and into the intestine. All this then comes out in the form of diarrhea, leaving the body malnourished and dehydrated.

Interestingly, there is variation across the global population in the number of TRs that people have. People with a greater number of TRs have a much harder time with cholera than do those with fewer. Those with fewer are less vulnerable to choleric diarrhea: they do not have as many regulators available to become bunged up or infected by the cholera toxin and break down in the first place. They therefore will not have as many TRs pulling fluids out of the body into the intestine and creating diarrhea and will not suffer as much as a person who has more regulators. However, although protected in this way against cholera, people with lower TR counts only will prosper in cool climates. If the climate is hot, they will suffer from other health challenges because of a related inability to sweat well to keep cool (the TRs implicated in cholera occur in places other than just the intestine).

It turns out that the TRs we are talking about here—which actually and more specifically are called CFTRs—are themselves regulated by a particular gene that has various alleles, one of which works much like the sickle cell allele that we learned about in Chapter 2. That is, it is partially expressed when inherited from one parent—and it confers an adaptive advantage when cholera is a threat (for the sickle cell allele, the equivalent disease is malaria). The allele's partial expression includes the generation of fewer CFTRs, so it makes its bearer less vulnerable to cholera. This allele is more likely to be found in people from parts of Europe where not being able to sweat very well generally poses little problem. The allele is thought to have evolved in these populations as part of a balanced polymorphism that protected the majority against death by cholera, albeit at the cost of sacrificing a few members of the population to the disease that inheriting two copies of the recessive gene would cause. That disease is cystic fibrosis (CF).

With some CFTRs missing, a heterozygous person living in an area prone to cholera outbreaks is advantaged. He or she has a lower chance of dying from cholera than does the person who has inherited two alleles for healthy CFTRs—the dominant homozygote. The recessive homozygote, however—the person who is homozygous with the recessive allele—thereby lacks healthy, functioning CFTRs. For this individual, real trouble ensues. Fully expressed, the recessive genetic disease cystic fibrosis disrupts the body's ability to absorb nutrients, breathe, and sweat. Cystic fibrosis was, prior to the development of modern treatment regimens, generally fatal in childhood.

The Germ's Point of View

I am telling this story through the disease ecology lens. The existence of this protective recessive allele reflects evolution in action. We made a change in our environment so that we were more likely to drink water containing excrement fouled by (among other things) the cholera bacteria. This change benefitted the cholera pathogen, but not us. For our population to survive, we had to change something in response to that. One kind of adaptation could have been cultural: we could have invented sanitation, as we eventually would do. Before we did so, however, genetic adaptation went to work. The gene pool happened to have a pre-existing variation in the polymorphic CFTR gene. One of its alleles became advantageous to carry when cholera came onto the scene. It was then naturally selected for.

These kinds of small, microevolutionary changes are happening all the time, in us and in the world around us. We see it now, for instance, in multi-drug resistant tuberculosis, wherein the tuberculosis or TB pathogen has evolved ways to resist being wiped out by particular pharmaceuticals. In a nutshell, the few TB germs that are resistant to a particular drug will survive it, reproduce, and repopulate with more resistant versions—versions now better fitted to survive in an environment flush with drugs that easily killed ancestral strains.

Not all evolution leads to increased or enhanced resistance. One of the ways pathogens evolve is simply to become able to infect a new host, for example through zoonosis. Another direction is toward **attenuation**: if a pathogen's impact upon its host becomes sufficiently minimized, hosts carry it around longer. If germs had intelligence, attenuation could be classed as a very smart move. The super-virulent disease that kills off all its hosts thereby sabotages its own existence; but a good host is a living host, so a well-adapted pathogen has a stable life with lots of living hosts to inhabit.

Many scientists think that the common cold was once a very virulent disease that attenuated; it became weaker in its effects and that was beneficial to the germ itself because it allowed the germ to spread much more easily. We see attenuation happening right now in a particular strain of cholera. Once people realized that water sanitation was key to stopping the spread of cholera, the classic strain was placed at a disadvantage. It also was disadvantaged by public health efforts to get people to stop eating ceviche and undercooked foods and encouraging the consumption of only boiled or disinfected water. Our cultural adaptations placed real environmental pressures on the cholera-causing pathogen. As a counter-response, classic cholera, which is very, very deadly, seems to have evolved one strain, the El Tor strain, which is better adapted to thrive within the confines of water sanitation.

El Tor causes less mortality and morbidity in humans, and as a result is excreted by human hosts for a much longer period of time (classic cholera victims die fast). Moreover, El Tor lasts longer outside of human beings when in water, and it can survive even in moderately cooked food. To top it off, people do not have any post-infection immunity after recovering from this new strain. In sum it does not make us as sick as classic cholera does, but it can survive a lot and infect more people. From a disease ecology point of view, El Tor is well-adapted to its present environment (Joralemon 2006, Chapter 3).

The Pivotal Role of Niche Construction

In the future, in addition to efforts to eradicate epidemic and pandemic health threats, we might further explore the possibility of culturally manipulating features of the environment that support attenuation in problematic pathogens. We also may be able to track, in real time, evolution not only in the pathogens that trouble us but also in the human genome. In either case, we should be wary of unanticipated knock-on effects. A small change in one area—in a gene, in a pharmaceutical concoction, in how we treat water or process waste, and so on—can have a huge impact elsewhere in our global system.

Through the cholera and the other examples in this chapter, we have seen that viewing human biology as static or unaffected by the environments that we have created to live in blinds us to the very real and very important connections between our physical bodies and the large, complex system that we are part of as inhabitants of the earth. We have seen how the various susceptibilities different populations have to particular diseases is a result of diverse interactions between the varied environments we have created (cultures) and the bodies (biologies) with which we inhabit them.

Dynamics of Human Biocultural Diversity: A Unified Approach by Elisa J. Sobo, 165–183.
© 2013 Left Coast Press, Inc. All rights reserved.

9

Political Economy of Variation in Human Health

This chapter prepares you to:

- Use the political economy perspective to explain the differentially distributed impact of social stratification on human health (epidemiological polarization)

- Describe the key components and main aim of primary health care (PHC) and demonstrate the value of PHC in combating epidemiological polarization

- Explain the need for a social justice approach to combat structural violence

In this chapter, using the political economy perspective, we examine the differentially distributed impact of social stratification on human health, as reflected in epidemiological profiles. An **epidemiological profile** is the disease profile or picture of a given group; it identifies what diseases and other health challenges they are experiencing or have experienced, and to what degree. It also attempts to say why.

Epidemiological profiles can shift over time. Recall that if a foraging group becomes agricultural, its epidemiological profile shifts to highlight epidemic diseases. What happens when agriculturalists intensify?

In the late 1800s, the Industrial Revolution was well underway in many until-then wholly agricultural societies. With it came not only the rise of factory-based industries but also the demise of ways of life that had, until then, been longstanding. People migrated to the cities in large numbers. Living conditions for workers were often unhygienic, uncomfortable, and crowded. This of course was a boon to the infectious diseases that thrive in such circumstances. However, as the industrial revolution became the industrial way of life, a new epidemiological profile emerged. In that profile, the diseases that bothered people most were not so much infectious diseases with high mortality rates; rather, they were non-infectious or chronic diseases with low mortality but high morbidity rates: people lived longer, but with longer lives came the diseases of old age.

This shift has been termed the **epidemiological transition**. To focus on this particular transition as 'the' transition of course ignores the importance of the forager-agriculturalist transition—but that is, anyhow, what epidemiologists have called it. As we shall see in this chapter, the simplicity implied in the label and its narrow focus conceal great complexity and a broader set of patterns related to human biocultural diversity—complexities and patterns we aim to explore.

Soap and Hot Water

What underwrote the epidemiological transition? In a word: sanitation. People began to wash their hands and take other steps such as building better sewerage systems. Why did they not wash their hands previously? For one thing, running water was not always easy to find; when it was available it often was cold—sometimes very. Moreover, people did not know that hands could carry pathogens from person to person.

One day, however, an obstetrician named Ignaz Semmelweis realized that the midwives in a hospital where he worked were having much better outcomes than the medical students. Semmelweis's data showed that the medical students were actually killing off birthing mothers because they were going straight from doing autopsies and working with cadavers to delivering babies, bringing germs with them. Even so, it took the medical community a long time to come around to Semmelweis's suggestion for hand-washing. Culturally, it did not make sense. The germ theory of disease had not yet gained credence, and people at that time did not know that microscopic pathogens could exist. People for the same reason did not think twice about drinking water if it looked clean enough, or about drinking from a shared cup.

Although we'd been living in our own germ-friendly filth for quite a long time, people in the later 1800s were forced to confront this because the factories that they built were so obviously polluting and tenement housing was so flagrantly awful. That's not to say that medieval cities were not redolent because they were—but by the late stages of the industrial revolution there had been developments in science that helped people see what they previously could not. Moreover, democratizing political changes

Figure 9.1
"Five cents a spot": unauthorized immigrant lodgings in a Bayard Steet tenement, New York. Photo by Jacob A. Riis, c. 1890.

and a belief in citizens' rights that had been underway already in Europe for a few hundred years now came to the fore, as seen for instance in the French Revolution. Friedrich Engels's classic 1844 study of the horrible conditions under which the lower working classes in Manchester England labored and the slums in which they lived spurred further activism, as did Jacob Riis's late 1800s photographs of squalid U.S. tenement life (Engels 1970; Riis 2004; and see Figure 9. 1).

The sanitation movement was really a social movement. Social justice campaigns that emerged out of the realization of how awful workers' lives were did a lot to clean up the cities and otherwise promote sanitation. Yes, the government was interested in stemming any potential for worker unrest that foul living conditions might inspire. Yes, capitalists were interested in keeping the cities attractive to rich shoppers. However, it was really the social justice campaigners who underwrote the epidemiological transition, and they did so in the name of the working class.

The Great Epi Divide

Polarized Profiles

Sanitation led to a shift in the overarching epidemiological profile away from high levels of mortality from infectious disease and lots of death at young ages toward a profile in which people lived longer and died later from non-infectious diseases. Only, not every population made the transition.

Many groups have maintained a pre-transition epidemiologic profile. These groups comprise whole countries—poor countries—as well as the poorer classes in rich countries. This **epidemiological polarization**—this division between the haves and the have-nots, or the transitioned and the non-transitioned—is called by some "The Great Epi Divide" (Kidder 2003, 125).

When we refer to biocultural diversity today as measured in terms of differential health outcomes and different epidemiological profiles, we are really talking about the consequences of holding different positions within local and global social structures. Health inequalities highlight parts or features of the social structure that divide societies like fault lines, or which exist as gaps—as places where people can fall through the cracks. Social stratification, which arose as a byproduct of the agricultural revolution, is a very important factor underwriting epidemiological polarization. Better off societies (such as U.S. society) have much better health outcomes; they have made the epidemiologic transition. Societies at the bottom of the scale (such as Haitian society) are not so well off health-wise (or any other wise).

We also see epidemiological polarization within societies. I am reminded of it every day when I see people waiting on a particular street corner in my town for someone to drive by and offer them work. Then there are workers who are exposed through their jobs to toxins and hazards. Agricultural workers in the United States have a very different health profile than, say, chief executive officers or CEOs.

Political Economy

In disease ecology, the focus is on evolution—in fact on co-evolution, as seen for example in the evolution of resistance in germs assaulted by culturally invented drugs. In disease ecology, we focus on **proximate** causes—for instance, factors in the immediate vicinity of infectious processes. Proximate variables are those things we see when we take a close up look.

On the other hand, in the framework offered by the **political economy** perspective, which we now shall adopt, you must look with a wide angle lens. This perspective considers phenomena or processes from a position much farther back. In political economy, rather than to focus on proximate factors, the focus is on **ultimate** causes—causes further removed but that precipitate an event. The political economist asks questions such as "Who benefits from this situation?" "Who is profiting?" or even "Who is just maintaining their gains, maintaining their position, or keeping the status quo as it is (that is, in their favor)?" The political economy perspective asks about power relationships; it asks about authority. Much of this authority is tied up of course with the economy—with money—and with whomever has it.

Multi-drug Resistant Tuberculosis

Tracy Kidder's 2003 biography of Paul Farmer illustrates, with numerous examples, the advantages offered by seeing biocultural diversity through the critical lens of political economy. Take, for instance, the case of multi-drug resistant tuberculosis, or MDR TB, in Russia. There, due to political and economic changes, many people have been thrown into poverty. Many have ended up in prison due to crimes committed in circumstances of severely limited access to legal, paid employment by which to feed one's family.

What happens in jail? People live in very close quarters where they are exposed to tuberculosis, a bacterial infection of the lungs that generally leads to death if untreated, but which, being slow-acting, has lots of opportunity to spread. Russian prisons simply have not been able to afford to provide sick prisoners with the right kinds of medical treatment on a consistent basis. Over time, a highly drug resistant strain of TB has evolved. Prisoners infected with this strain can spread it to prison staff members and, upon release, outsiders. This is one reason that Farmer's colleague, Alex Goldfarb, calls prison an "epidemiological pump" for the MDR TB crisis (Kidder 2003, 232).

When we consider TB from a political economy point of view, we do not ask so much about what a TB germ does once it has entered a person's body or how it is spread per se; that is the purview of disease ecology. When we look at things from a political economy perspective, we ask who is exposed, why they are exposed, and why they are vulnerable to a problem anyhow. What in their political-economic or social context set them up for infection? With TB, for example, we ask who is at risk for incarceration. We ask who is malnourished, too, because the better nourished people are, the less vulnerable they are to infection. We ask about who has access to the right medicine, and why. We ask why one class of people get TB and another does not.

I mentioned that the prisons do not have the money to buy the right kind of medicine for incarcerated patients. "We just don't have the money" is, as Farmer says, a very common phrase in explanations of health inequities. What Farmer shows and what many theorists also argue is that "we" do have the money. We just prefer to spend it on other things—on luxuries for our pets for example, or fancy cars, or status handbags, or any number of non-essential consumer goods. There is in fact enough money circulating in the world. The problem, then, is in its unequal distribution; it is that priorities for spending do not extend to the provision of appropriate pharmaceuticals to all who need them (Kidder 2003; see also Farmer 2005).

Structural Violence

Sometimes certain social groups are cut off from access to health care or put into positions that make them vulnerable to exposure, such as to TB. Sometimes they are put at risk or in certain conditions, such as via the occupational hazard of inhaling pesticides, that can lead to cancer and so forth. The harm brought on by this kind of vulnerability is a form of what Johan Galtung calls "structural violence" (1969). In **structural violence**, the shape of a given social structure harms or is harmful to the people who occupy certain positions within that social structure.

Some social structures cannibalize themselves, using up their own human resources in structurally violent acts. This is quite obvious when we consider a cotton or sugar plantation fueled by slaves—individuals treated as human resources: purchased, and easy to replace when they burn out or break. In reality, of course, slaves are people—people who, because of their position in the social structure, come to harm. They are consumed by the system as it strives to keep itself going.

Before going on, let me make sure that we are all clear on exactly what social structure actually means. Societies are made of groups of people, and all of the people in a society have some kind of a relationship with one another. The **social structure** emerges from those relationships, just as a building can emerge out of the relationship of wood, nails, and concrete. Put together one way, a ranch house gets built; another way, an A-frame cabin; still another way, a two-story abode (see Perry 2003, Chapter 4). Each structure emerges out of the relationships between the bits of building material. Likewise, if we fit people together in one way, one kind of structure will emerge; if we fit them together in another, we can have a different social structure entirely. Think of the difference between an egalitarian social structure and, say, a feudal society governed by a monarchy. Part of the difference lies in the impact of each social structure on the well-being of particular subgroups, which in turn depends upon how its pieces—its subgroups—are related.

In the slave-fueled plantation, the impact on slaves was quite obvious, which is why I began with that example. However, structural violence (like any effect of the social structure) generally seems quite unremarkable. It generally remains unseen because it is understood as part of the natural order. In this way, slave deaths and maltreatment were passed over by members of slave-owning societies as nothing problematic. Likewise, in our own society, there are rhetorical tricks that we can play to keep ultimate causes for certain groups' suffering hidden. For instance, we might write people's problems off to just plain bad luck or individual, proximate factors rather than pulling back to get a wider-angled view of the structurally promoted aspects of the individual's situation.

Structural violence, then, describes the hidden violence entailed by a given social structure. There are lots of kinds of violence that we cannot see with the naked eye. There are long term insults to the body that build up over time, for instance. One way to think about this violence is to consider that each individual does bear some of the burden of supporting his or her society, whatever its shape. Wherever one is in the social structure, one must bear some weight. How much weight one bears and where it is borne on the person will differ depending upon where a given individual is located in the social structure. Moreover, people who are further down in a social hierarchy have much less **agency**—they have much less ability to impose their will or make choices than the people who are at the top. To add insult to injury, they are not even given the same kinds of choices to make as people at the top of the heap.

In structural violence, the forces of history come together to structure a society in a particular way, and link it up with other societies in a particular way. If we think back to the changes brought on by the rise of agriculture, and work our way forward through history until the present day, we can see that most structural arrangements track back to economic issues emerging from the organization of subsistence activities. All the historical economic forces that are bundled up in a social structure get written into the bodies of the people who occupy different positions within the social structure in diverse ways depending upon the position occupied (note that here 'bodies' includes minds, despite the dualism that the English language tries to force upon us; for more on this, see Chapter 10).

While there certainly are some costs to the people on the top of any social structure, it is at the lower levels that inequity is most pathogenic. Moreover, it is not just class that is implicated. Gender, too, has long been a part of this process; and as societies became more heterogeneous, ethnicity and so-called race also entered the equation. The more complex a social structure is, the more complicated and even convoluted structural violence can be. Just pointing to structural violence generically as a cause of harm, then, is not enough; the successful application of the concept demands specifics regarding hows, whats, whens, whys, and so on.

Structural violence is not just a local phenomenon; frequently it reaches around the world. Each social structure exists within a larger one, and poverty in one location in the world actually can support wealth in other places. This is the case, for instance, in the Haitian countryside area where a dam (the Peligre Dam) was built to power foreign-owned industry in the capital city. The peasants living where the dam was to be built were not consulted, and in the end they were forced to flee when seemingly all of a sudden water began to rise and cover their homes and gardens—all for the short term benefit of Haiti's foreign investors and urban elite (Kidder 2003).

Losing one's home, getting tuberculosis, getting a dental infection and losing one's teeth due to a lack of affordable dental care, having a hand chopped off in a meat grinder at work that had no proper safety mechanisms—these kinds of facts represent more than just individual bad luck. Such events do not happen randomly; they happen to particular people because of the positions those people occupy in the social structure. Now this is not to say that individuals should not take responsibility for themselves. Of course they should. However, many occupational hazards exist not because of worker actions but because employers are not protecting workers. Even when a person working two jobs has an accident at work due to tiredness, something else also is to blame—and that is the system that makes working two jobs necessary for this person to get by.

Structural Violence in Historical and Contemporary Perspective

Structural violence is not a new phenomenon; it has been around since the rise of agriculture at least. We know that, compared to groups lower on the intensification scale, agricultural groups are larger and more stratified. They have more inequality. With more inequality we see different subgroups being exposed at different rates to different hazards. We see different subgroups being allowed at different rates to access healthcare and healthy lifestyles. We see this in the archaeological evidence.

We can see 10,000 years of health disparities written into the archaeological record if, for instance, we look at what skeletal and material remains reveal. In Chapter 7 we considered some of the signs of non-specific nutritional distress. While pervasive in whole societies after agriculture was adopted, such signs were more pervasive in some subgroups than others. Some archaeological sites include separate burial grounds, and we can see from the skeletons of the elite how well-nourished they were in contrast to the working class or the slaves. We can also see different types of occupational stress and injury on the bones that we find. These data show that health inequity related to social structure has been around for a long, long time.

Health inequity is ongoing now, too. We have masses of data on class-based inequities in many nations. In the United States, much inequity shows up along so-called racial or ethnic lines. For instance, in terms of deaths due to diabetes, we have much higher rates among Hispanics and blacks than whites. The difference is actually getting bigger—and it is not the only health problem that plagues blacks and Hispanics more than whites. There are many, and often they feed each other in a synergistic fashion.

Syndemics

This is my cue to add one last word to our epidemiological vocabulary, and that word is syndemic. We've had endemic, epidemic, pandemic; a **syndemic**, in contrast, entails not just one health problem but a cluster of problems that work together, reinforcing and often exacerbating each other synergistically. One example of a syndemic would be HIV disease,

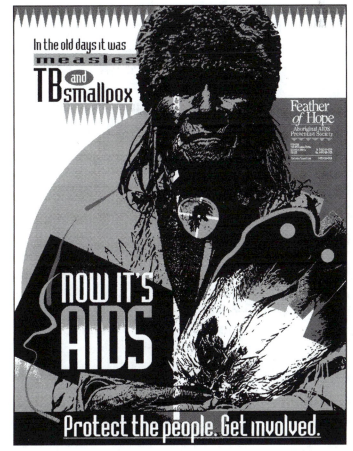

Figure 9.2 Poster from the Aboriginal AIDS Prevention Society illustrating persistent and syndemic health problems. *Image from the History of Medicine collection, National Library of Medicine, National Institutes of Health, 101451231.*

or AIDS, occurring hand-in-hand with TB, and both of those reinforcing each other in a given population. Moreover, where we have a very malnourished population, we will have people at a higher risk for being infected with HIV and TB.

What can we do when we see a syndemic? We can help in the immediate situation. We can provide drugs, bandages, stitches, and nutritional supplements, for instance. Yet as Albert Einstein is credited with having said, "No problem can be solved from the same level of consciousness that created it." We have to step back from proximate concerns to try to understand the situation in its broadest context because until we understand and address the ultimate causes of the situation, the facts that have created it will not change and it will remain an ever present problem (see Figure 9.2). The only real way to prevent or control a syndemic is to prevent or control the forces that lead to that cluster to begin with. Again, those forces are political and economic; they have to do with power relations—with relations having to do with money and its circulation. Those relations must be altered to prevent or control syndemics.

Sustainable Health

Health is an indicator of how well populations are adapted to the environments in which they live. While the human species may seem to be thriving, the evidence actually shows us that it is not fully optimizing its potential—at least not across the board. Great diversity can be seen when we compare the epidemiological profiles of various groups.

In many cases, people are struggling; the systems in which they live are generating problems for them—problems expressed in the form of poor health outcomes. Variation in the distribution of these problems is tied to variation in living conditions. Variation here is, in turn, tied to the political economy—both the local political economy and the global political economy.

The existing political economy was humanly created. Humans created the global systems that exist. If humans created them, humans can change them. This may sound daunting, but as Franz Boas's eminent student Margaret Mead is famously claimed to have said, "Never doubt that a small group of thoughtful, committed people can change the world. Indeed, it's the only thing that ever has."

So far, two approaches to health improvement that have been very successful are primary health care and the social justice approach, both of which we are about to discuss. Used in combination, both approaches are very valuable in combating epidemiological polarization and the structural violence that sometimes brings it about. They are valuable in combating the negative kind of biocultural diversity (that is, in health status and outcomes) so that the positive kind can flourish.

Both approaches have their roots in health programs that were instituted after World War II when the United States got heavily involved in overseas aid provision. Through organizations like the U.S. Agency for International Development (USAID), U.S. experts and staff members conducted many health programs all around the world.

Initially, most such programs failed. As we came into late '60s and '70s, people started to wonder why, with so much expertise, and money, the programs were not working. Anthropologists and other social scientists were called in by the government to study the situation. They concluded that the programs were failing because they were imposed from the top down. There was no community involvement and no cultural connectivity.

Social Soundness

In 1975 USAID instituted an initiative called **social soundness**. This initiative called on all aid groups to take up the specific goal of community involvement—of connecting with community leaders and members and working to create a good fit between programs and cultures. At the same time, leaders and aid workers all around the world were noticing that many of the health issues that they were trying to address stemmed from poverty. They realized that these problems would not actually abate until poverty itself was alleviated. They saw that in many cases solutions were technologically simple, and cheap. They entailed, for instance, providing clean drinking water, immunizations, and mosquito nets.

Such solutions sat at the core of a very important meeting convened in 1978 by the World Health Organization (WHO) in Alma Ata, Kazakhstan—a meeting attended by people, including leaders, from all over the world. At this meeting, people declared that we would achieve health for all by the year 2000. This led to an initiative that we still have in place today, although its name has been updated to "Healthy People 2020." The goal is to achieve health through simple solutions. One of those simple solutions is primary health care.

Primary Health Care

The prime mover of the epidemiological transition, where it has happened, was *sanitation*. The prime mover in the primary health care approach is *prevention*, or preventing problems before they happen. Key routes to prevention involve water, food, and adaptive immunity.

Primary health care first entails providing people with safe drinking water. Even in the United States, assurance of safe drinking water was in

Figure 9.3 International Medical Corps doctor examines a patient at a mobile health clinic in Pakistan. *UK Department for International Development via Wikimedia Commons, CC-BY-2.0.*

the 1970s a new thing: the Safe Drinking Water Act of 1974 was brand new when people met at Alma Ata to begin to define this new direction.

Nutrition is the second component of the primary health care approach. This does not mean dropping famine relief supplies from a plane. It means helping people to achieve a sustainable food supply.

The third component in the primary health care approach is immunization. This helps ensure that people do not get sick in the first place.

Some proponents of the primary health care approach go even further than to champion prevention via clean water, good nutrition, and immunization. For instance, Partners in Health, a non-governmental organization dedicated to serving the poor, also recommends free health care and education for the poor, community partnerships that address basic social and economic needs, and that we serve the poor through the public sector. This is exactly the kind of approach that many people talked about in Alma Ata (see Figure 9.3).

Social Justice

Partners in Health adopts a **social justice** approach, which promotes equitable distribution of basic human rights such as the right to healthful living conditions. Social justice proponents promote equal opportunities for equal outcomes among all social groups so that one population is no more privileged than others. There are many groups in dire need of social justice—not just in poor nations but in poor segments of our own country.

Distribution of Lead Poisoning

Figure 9.4 shows demographic and epidemiological data from New York City (the racial and ethnic categories the city used conform to U.S. census norms). The first bar in each set represents the percentage of all children under eighteen years of age who belong to the particular ethnic or racial category listed. The second bar represents the percentage of lead poisoning cases among children when sorted into the same ethnic and racial groupings. These are cases that required environmental intervention, which means clearing lead paint and lead pipes and so forth from their houses. Recall from Chapter 7 that lead poisoning in children interferes with nervous system development, and so can cause developmental problems—if not death.

Notice that the bars of each type are of different heights for most groups. For example, while 9 percent of New York's child population is Asian, Asian children account for 18 percent of its lead poisoning cases. Likewise, while 29 percent of the children are black, black children account for 42 percent of those poisoned by lead. Conversely, while 24 percent of children in New York City are white, white children account for only 5 percent of those suffering from lead poisoning. The distribution of lead poisoning by racial and ethnic group does not match the distribution of race and ethnicity in the population at large.

To try and answer the question of why not, maps can come in handy. Had we a map of New York City showing where poverty is concentrated,

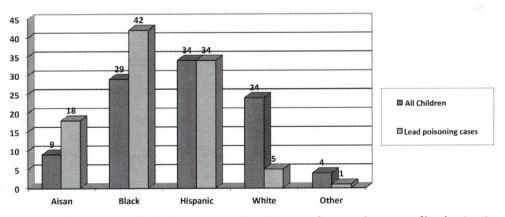

Figure 9.4 Percentage of children younger than 18 years of age, and percent of lead poisoning cases among children that required environmental intervention; New York City, 2004. *Created by author from data shown in Mullin 2004.*

and a map showing where the identified groups are concentrated, and a map showing where lead poisoning cases are concentrated, we would easily come to the conclusion that poverty and lead poisoning are positively correlated, as are poverty and race or ethnicity. If someone lives in a poor, minority area, he or she is much more likely to be exposed to lead. The housing stock is old. It is not well taken care of: nobody repaints the houses; nobody replaces the pipes. This means that children in these areas, who already are disadvantaged in terms of their impoverishment and other factors, such as malnourishment, poorly performing schools, and so forth, now have the added disadvantage of lead exposure on top of everything else.

Parasite Stress, Iodine, and IQ

Many things beside lead can dampen children's developmental potential. Whatever problems there may be with defining and measuring human intelligence quotients or IQs, the synergistic link between higher disease burdens (including malnutrition) in particular populations and lower IQ scores in those populations is by now well established. We know less regarding exact causal mechanisms and patterns of, and causes for, intra-group variation and resilience. Still, recent findings concerning the correlation between IQ and life years lost to twenty-eight infectious diseases for one hundred and ninety-two countries is indicative. This study tested the "parasite-stress hypothesis," which holds that fighting a high burden of childhood infections caused by viruses, bacteria, and things like tapeworms that live, parasitically, on the human body can be such a stress—can divert so much of a child's energy—that normal cognitive development cannot take place.

If the hypothesis were wrong, you might expect no correlation between IQ and the disease burden index the researchers created. Yet a strong inverse correlation did exist. The highest average IQ measured was in Singapore (108), followed by South Korea (106). Third place was a tie, between Japan and China (105). Of English-speaking nations, both Australia and the United States averaged 98, Britain averaged 100, and Canada 99. All of these nations have low levels of infectious disease, relative to those whose IQ averages were lower—and the lowest was 59. Sure enough, the disease burden for that lowest group was 4.45 as per the study index; the highest group's, Singapore's, was 2.67 (*Economist* 2010a; Eppig, Fincher et al. 2010).

Even the simple lack of access to iodine is associated with impaired cognitive development. Most of us get plenty of this micro-nutrient via iodized salt, but about one-third of the global population does not get enough of it. The implications are enormous: "When a pregnant woman doesn't have enough iodine in her body, her child may suffer irreversible brain damage and could have an I.Q. that is 10 to 15 points lower than it would otherwise be" (Kristof 2008).

Knowing what we know now about the (negative) synergy between malnutrition, disease burden, exposure to toxins, and so forth, and IQ, the word 'syndemic' should be coming to mind. Importantly, then—and even leaving out the impact of culture on IQ score (which can be great; for instance those who do not come from the culture that devised a given test are unlikely to score high on it)—we can see clearly that diversity between populations in terms of measured IQ averages is humanly created.

Put another way, reports that people living in certain areas or people who are members of certain groups have "lower IQs" on average than people from other areas are, in fact, true—but they do not mean that populations are inherently more or less clever. When put into context, they only show that population-to-population IQ differences, like so many aspects of diversity within the human species, are to a large degree humanly created. This kind of diversity is therefore not at all a good candidate for being used as a 'natural' justification for oppressing certain populations or limiting their access to, say, education—although those who are ignorant about the link between IQ and environment sometimes claim it to be so.

Zoning

Back in New York, poor people's overexposure to environmental toxins (and so forth), and the ramifications thereof, has much to do with their poverty. It is therefore a product of the political economy that demands that some people subsist as low-wage laborers if others are to amass wealth. It also stems from the political economy's related support for zoning laws that allow environmentally problematic businesses—which would not be allowed in 'better' neighborhoods—to operate in poor ones.

For instance, zoning laws permitted metal stripping and plating facilities to exist in close proximity to residential housing in the Barrio Logan section of San Diego. This neighborhood is home to many lower income Latino families. Yet plating operations release toxins, including heavy metals, into the air. Those who live near them are thereby exposed. Plus, a lot of toxins go home on the clothes and in the hair of people who work at such places, meaning that workers' family members, too, are exposed to the toxins worked with.

Barrio Logan's residents were used to the plating facilities, but after awhile they noticed that they seemed to be experiencing higher rates of miscarriage, breathing problems, headaches, and other physical ailments than people in nearby neighborhoods without such facilities in them. Residents got together and, with the support of some people who helped them collect data, they secured an intervention from the city, changed some laws, and got the plating facilities to close down.

Of course, similar stories do not always have such happy endings. Plus, environmental pollutants from the plating facilities were not and are not the only problems that Barrio Logan's residents have faced. They and residents further south also have been exposed to chemicals coming from the shipbuilding yards that are close by. Similar to the situation in New York, in Barrio Logan the housing stock is old, and so residents have been exposed to lead and asbestos (a known carcinogen). There also are many 'brownfield' sites there—sites where contaminating activities have taken place in the past and which now host schools or other facilities, sometimes without having undergone the environmental remediation necessary to make the site a safe and healthy place to be. Proponents of the social justice approach find this way of treating the poor unconscionable. Be that as it may, the examples above make the role of the cultural side of our biocultural existence—of economic forces and social and political decisions—manifestly evident.

Community Driven, Culturally Responsive Change

To close this chapter, let us look to a case study offered by the Tohono O'odham Nation of Arizona, who presently have one of the highest rates of diabetes in the world. This case study highlights the human creation of diversity, here in obesity and diabetes rates. More than that, it demonstrates the links between social justice and health while charting the ramifications of structural violence for the Tohono O'odham people; showcases the enactment of social soundness, as per Alma Ata; and exemplifies the importance of prevention, as promoted in the primary healthcare approach. Importantly, too, the programs that ultimately may heal the Tohono O'odham Nation emerged from within that group—not from without. They are truly community based.

Several generations ago the Tohono O'odham lost their right to the region's water when the river fueling the farming lifestyle typical in the area at the time (see Figure 9.5) was diverted to service nonnative farmers upstream. At that time, and in fact until the 1960s, no known case of diabetes existed among the Tohono O'odham. However, when they lost their farm-based livelihood and the cultural ways that went with it, they also lost the means to good health: physical activity, a balanced diet, and a hopeful outlook (California Newsreel 2008a).

Figure 9.5   Tohono O'odham woman winnowing (cleaning) wheat c. 1907. *Library of Congress, Prints & Photographs Division, Edward S. Curtis Collection, LC-USZ62-123312.*

Today, more than half of all Tohono O'odham adults have been diagnosed with type 2 (adult onset) diabetes—as have increasing numbers of Tohono O'odham children. Because of their unnaturally elevated diabetes rates, the Tohono O'odham have become favorites for biomedical research. However, rather than to simply submit to genetic testing in support of the biomedical community's individualized, gene-based understanding of why they are more likely to have diabetes than other populations, or to otherwise offer themselves as human test subjects for others' research projects, the Tohono O'odham Nation came together to form Tohono O'odham Community Action, or TOCA, a community-based organization dedicated to re-creating a sustainable, healthy culture (see Tohono O'odham Community Action 2010).

TOCA promotes cultural renewal, and the empowerment this entails, as key to wellness. At present, TOCA has four primary program areas: basket weaving, traditional foods, youth and elder outreach, and traditional arts and culture. All are intertwined. The traditional foods program area, however, is perhaps most directly related to diabetes abatement. It emphasizes healthy eating and seeks to re-teach traditional farming skills, which were lost to many when the river was diverted. This it does through school programs and other outreach efforts. Concurrently, TOCA strives to revitalize the cultural aspects of Tohono O'odham foodways, using native traditions in the context in which they were meant to be used rather than as, for instance, performing a ground-blessing ceremony on a stage. TOCA also helps to support traditional farming efforts by creating markets for the foods so grown, including via its website (www.tocaon-line.org/Home.html).

Rather than simply endorsing and working to ensure access to the current clinical model of diabetes care, which of course is important, TOCA emphasizes a community-based model of holistic diabetes intervention. It celebrates the community's farming-based roots, all the while casting the community as part of a larger subsistence system. Ultimately, TOCA aims to address the structural causes of unemployment and poverty and so of ill health among the Tohono O'odham people. Advocacy of the kind that led to the Arizona Water Rights Settlement Act, which went into effect in 2008 but was over thirty years in the making (Archibold 2008), has been a crucial part of this process.

Populations Embody Their Social Positions

We've seen in this chapter how power relations within and between social groups underwrite biocultural differences. In complex societies, and in the global political economy, a group's structural position has consequences—consequences that sometimes last into succeeding generations, affecting life chances (as well as sometimes genomes and epigenomes). As Nancy Kreiger says in the film *Unnatural Causes: In Sickness and in Wealth* (California Newsreel 2008b):

> There's one view of us as biological creatures; we are determined by our genes [and] what we see in our biology is innately us, [it is] who we were born to be. What that misses is that we grow up and develop.… We interact constantly with the world in which we are engaged. That's the way in which the biology actually happens. We carry our history in our bodies."

As this chapter has shown, structural position is part of the history that is so inscribed; it is part of what underlies our biocultural diversity.

Dynamics of Human Biocultural Diversity: A Unified Approach by Elisa J. Sobo, 185–202.

Part III

Meaningful Practices

In the first part of this book, we learned about and applied systems thinking to understand bodily adaptation to the physical world, both at the population level, via genetics, and at the individual level, via developmental plasticity or epigenetic mechanisms. We saw, too, how individual developmental adjustment and epigenetic change, on the one hand, and population level change, on the other, actually can be quite connected.

Once we came to understand how humankind emerged over time well-fitted to East Africa's physical geography, we examined what happened when people spread out. Still using the complex adaptive systems framework, we focused in Part II on the social structures humans created as they sought to exploit new areas of the world, and to do so in new ways, such as via settled agriculture.

While society (or social structure) and culture are two sides of the same coin—with 'society' describing the relationships between available roles, and 'culture' justifying those relationships and roles—our focus has until now been on the social side of things. We did talk about culture, but only in terms of what it contributes toward human survival directly, such as in providing means for provisioning food, clothing, and shelter. The last few chapters primarily focused on the social systems that emerged.

We did not explore the cultural knowledge or belief systems that, for instance, support paying tribute to a chief of divine lineage, or submitting to feudalism, agreeing to rule a nation state, or serving as a shaman. We did not explore the ways in which people, however organized, attempt to manipulate supernatural forces to bring on better harvests, ensure procreative fertility, pull down a haughty or competitive neighbor, or encourage love. We did not examine various cultural knowledge and practices relating to kinship, gender, sexuality, age, or other attributes that can be used to differentiate groups of people.

Yet all these things entail biological and cultural interaction. The more cultural side of things is limited to some degree by the biologies we are endowed with, but at the same time our biologies reflect cultural actions—actions taken (consciously or not) in response to cultural meaning systems. In some instances culture creates the biology it desires; in others, effects on biology are secondary or unanticipated. Sometimes meaning systems are so deeply internalized that our bodies respond in ways that are beyond our awareness.

Chapter by Chapter Overview

Chapter 10 deals with meaning directly as it characterizes the link between social structure, mind, and body through a discussion of population-level variations in stress and its long-term effects. Stress is meaning-laden; different events may be perceived as stressful or not in different cultures. Indeed, different cultures may entail different levels of stress altogether. Studying the biocultural diversity this entails has important implications for health and well-being.

The link between culture, mind, and body is further explored in this chapter through discussion of circumscribed health conditions that occur in one culture but not another, and the role that culture plays in interpreting, experiencing, treating, and even generating particular symptom sets or diseases, such as *susto*, a form of soul loss known in certain Latin American populations. The diverse beliefs various populations hold underwrite diversity in self-healing capacities, as this chapter's exploration of the placebo effect reveals.

Chapter 11 encourages readers to explore the role of day-to-day, belief-fueled cultural practices in creating forms of diversity thought to be 'natural,' such as gender-related differences in comportment. To help grasp this, we consider what happens when people act in ways that can be classed neither as male nor as female, and what cultural features underlie such behavior. We then characterize and apply the Rites of Passage framework to demonstrate how expectations for manhood and woman-hood are culturally constructed.

Chapter 12 extends the analysis to include body ideals, with an eye to their biological impact. We explore cross-cultural and historic differences in weight ideals in relation to food supply, and examine weight-related health challenges in light of our evolved metabolic biology. Then we identify and describe the functions of body modification and decoration cross-culturally, such as through tattooing, and discuss the main dimensions along which such practices vary.

Beyond how people are shaped, who one's relatives are can differ cross-culturally. Chapter 13 describes the three most basic kinds of kin-ship, paying particular attention to cross-cultural variation in the ways humans create consubstantiality—that is, kinship due to shared sub-stances, of which blood is only one. The concrete, physical consequences that ideas about kinship have for family-building are examined. We also explore diverse understandings of when, how, and if life begins, and exam-ine how these various understandings affect practices such as abortion.

Central Lessons of Part III

In sum, Part III of the book pushes us to give equal weight to the cultural end of the biocultural equation by taking into account the impact that the deepest held ideas of different populations have for human biocultural variation. It guards us from settling for a focus that is solely material by demonstrating the value of a perspective equally sensitive to ideas and meanings. Having done that, the book concludes by reaffirming that biol-ogy and culture co-create each other and the human experience; holism and systems thinking are useful; variation is good; and we are all connect-ed. Areas ripe for future inquiry also are noted.

10

Diversity in Mind-Body Relations
Stress, Meaning, and Health

This chapter prepares you to:

- Characterize the link between social structure, mind, and body through a discussion of population-level variations in allostatic load (chronic stress) and its effects

- Characterize the link between culture, mind, and body through discussion of diverse culture-bound syndromes

- Explain how cultural models are used in interpreting, experiencing, treating, and even generating certain disease states

- Characterize the link between culture, mind, and body through discussion of the Meaning Response in terms of self healing

The last set of chapters concentrated on the relationship between biology and culture by highlighting how the various ways that groups of us humans make our livings, and organize ourselves, feed our biocultural diversity. Beginning with this chapter, our inquiry broadens to include, more directly, cultural meaning systems. Without losing sight of the co-determinate nature of biology and culture, we will first explore chronic stress and its diversely distributed causes and effects. Then, we'll take on culture-specific sicknesses and self-healing.

The Stress Response

What Is Stress?

Technically speaking, **stress** is one's immediate physical response to environmental pressures. The stress response entails reactions of the nervous system, such as when the brain releases certain neurotransmitters; the hormonal system, which releases stress hormones; and the immune system, in which some parts go on hold so that energy can be diverted to protect oneself from whatever the stressor may be.

It may sound surprising that hormones are involved in the stress response, because so often we think of them only in association with masculinization or feminization. However, they serve a broader function than just that. Hormones regulate the relationship between the various organs of the body. When a person is under stress, those relationships must change, because the stressor demands a different kind of response from those organs than every-day living demands. For example, exams are big stressors for students and teachers alike. Other stressors might include the threat of a layoff at work or that a snowstorm will close the highway one has to take to get home. Being attacked by a lion is stressful, too, as is undergoing a famine or being in a large earthquake. In each case, the body is mobilized to act quite differently than it does under normal conditions, and hormones play a role in fostering that mobilization.

Selection for a Strong Stress Response (Fight or Flight Reaction)

Why does the stress response exist? As humankind evolved over time, it was selected for because it was very helpful. The **fight or flight reaction**, which prepares the stressed individual immediately, via internal changes, for a fight or for fleeing a stressful scene, was an aid to survival.

Imagine a forager, quietly going about his or her business, inadvertently disturbing a rattlesnake. If he or she can act faster than normal in an effort to escape danger—which is what happens if the stress response is triggered—he or she is at an adaptive advantage. That is, because this forager survives when others would not, he or she is more likely to have more offspring, thereby contributing more of the genes that supported the stronger, quicker response to the next generation's gene pool. The frequency of genes supporting a stronger fight or flight reaction increases.

The Biology behind the Supercharge

Biologically speaking, what happens to cause an individual's body to quickly provide the energy needed to climb up a tree or run away or

Figure 10.1   Adrenal glands, located on top of the kidneys (as seen from behind). Based on lithograph plate from Gray's Anatomy of the Human Body, 1919. *Roxbury-de via Wikimedia Commons.*

otherwise do something out of the ordinary that will help her or him to survive? What causes the supercharge? The mind registers the threat and the body responds, churning out the necessary biochemicals, including hormones and neurotransmitters. One biochemical central to the stress response is cortisol. There also is a key category of biochemicals called catecholamines, which includes what used to be called adrenaline, now referred to as epinephrine. Epinephrine actually is generated in the adrenal glands, which sit on top of the kidneys (see Figure 10.1).

Exactly what happens when someone's cortisol level and so on shoot up? We all know something of what happens because we all have been there. Each and every reader has likely experienced the fight or flight syndrome (see Figure 10.2). From my own life, a very particular scenario stands out. I was writing a book (not this one) and had just finished Chapter 3. I began tidying my desk and then, in a wicked instant, realized that I'd just deleted the entire chapter. I later recovered it, but within that first moment of disaster I instantly felt as if a quart of stress hormones was coursing through my veins. My heart rate went up, my blood pressure surged, and sugars in my bloodstream mobilized, potentially providing my muscles a source of instant energy. Indeed, blood rushed to my muscles. This would have been terribly handy had I needed to run away from a bear or fight off an intruder.

Sadly, no such action could be taken. I could not run up a hill or climb a tree—well I could, but it would have been pointless because the stressor was not something I could counteract by doing that. My body simply had to reabsorb everything (I did go to the gym later, which provided some relief).

Figure 10.2 In situations defined or perceived of as stressful, some body processes speed up and others slow down, as indicated, in response to changes in neurotransmitter, hormonal, and other physiological signaling systems.

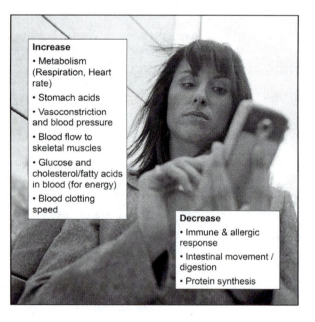

Increase
• Metabolism (Respiration, Heart rate)
• Stomach acids
• Vasoconstriction and blood pressure
• Blood flow to skeletal muscles
• Glucose and cholesterol/fatty acids in blood (for energy)
• Blood clotting speed

Decrease
• Immune & allergic response
• Intestinal movement / digestion
• Protein synthesis

How Stress Varies

As in my experience, most of us these days do not find ourselves stressed by challenges such as meeting up with grizzly bears or lions. But we still encounter similar stresses, such as when a car pulls out in front of us quickly, without warning. This kind of stress is, like an illness that hits swiftly, acute. However, many contemporary stressors are chronic. They persist over time. Even my assumed loss of that chapter was part of a chronic stress onslaught having to do with professional expectations. The contrast between chronic and acute is thereby a question of the level of analysis; and more simply, it is a question of duration.

Another variable is the source of the stress. It may be a permanent fixture on the landscape or it may be removable through certain actions. It may be work stress that one is powerless to do anything about, for example, or it may be a bear that actually can be shot. In other words, stressors can be more or less constant, meaning that in some cases the stress response is always set to on.

Chronic Stress: Allostatic Load

Given that individuals can and do respond to stressors with varying amounts of resilience, and that habituation is a possibility, generally speaking, when a stressor is constant and irremovable, a moment will come when the body gets so overloaded by its own stress response that it can in fact begin to break down, or undergo system failure. The fight or flight reaction that is so helpful when we really do have to either fight or run away fast can backfire or become dangerous when unabated.

One way to think about this is to picture a dump truck or trash chute that just keeps spilling garbage on a group of people. One dump is stressful enough, but dumping that does not stop—chronic stress—is downright debilitating. The process by which stress does its damage is not as simple as being crushed by garbage's weight. Rather, it has to do with recalibrating bodily set points to try to stave off system failures that the onslaught of stress otherwise might bring on. This onslaught also is known as the allostatic load.

To quote an expert, **allostatic load** is the "cumulative multi-system physiological dysregulation that results from exposure to challenges over the life course and that places individuals at greater risk for poor health" (Gersten 2008, 532). Dysregulation is when our normal regulatory processes (for example, metabolism of sugar) are disturbed or discombobulated. When a particular habitat includes certain long-term environmental pressures or chronic stressors, groups of people are affected by a high allostatic load.

A heightened allostatic load can lead to long-term immune suppression, infertility, atherosclerosis or hardening of the arteries, and heart disease among other things. It can promote abdominal obesity by affecting the pattern in which the body stores fat. When fat accumulates in the gut, around the organs, rather than in the buttocks and thighs, it is more likely to harm the vital organs it thereby surrounds. In addition, coupled with the fact that people under chronic stress have extra sugar available in their blood, it can underwrite the development of diabetes. Hypertension is also a possible outcome. In these ways, chronic stress writes or inscribes itself into the metabolic and cardio-pulmonary biology of a given population.

Chronic stress also seems to tax the brain. It may even do so in a sinister way that perpetuates cycles of poverty. Working memory—the ability to hold informational bits in your brain for immediate use—seems to weaken in people under stress. Moreover, if allostatic loads are high in childhood, which they are for children living in poverty, adult brain function can be affected. Adults who grew up in poverty are able to hold fewer items in their working memories at any one point in time than adults raised in less stressful circumstances. On top of that, because poor (stressed) children's memories do not therefore work as well as their less-stressed middle and upper class counterparts, they get lower marks in school, end up similarly poor as adults, and raise their own children in likewise poor circumstances. The synergy between nature and nurture here is clear: brain biology helps determine social status and social status helps determine brain biology (*Economist* 2009).

Of course, memory problems and disease responses such as in diabetes are exacerbated by other contributing factors, such as lack of sleep due to working two jobs or crowded living conditions, or poor diet. Nevertheless, from a medical viewpoint, they stem directly from the recalibration of once-healthy set points to allow stress-induced levels of certain biochemicals or responses to persist.

In fact, scientists named ongoing stress the allostatic load in reference to **allostasis**: the process of re-creating what we've learned to call homeostasis by changing the body's initial set points or 'factory settings' in order to accommodate chronic stress (as in allocare, 'allo' means other). Regaining balance in this way is good in the short term because the failure to maintain homeostasis itself can cause death. Ultimately, however, if the new balance is unhealthy, long term damage comes about.

Accordingly, common measures of allostatic load also are indicators of long term damage. They include high blood pressure, a large waist-to-hip ratio, high urinary epinephrine and cortisol levels, and so on. Such indicator scores generally are combined in studies to create what's called an 'allostatic index.'

New science regarding the allostatic mechanisms by which stress does its damage is both fascinating and groundbreaking. However, the link between stress and ill health has long been known. In the American context (we will get to some others later), for years studies have demonstrated a positive correlation between poor health outcomes and household stress levels (indexed by measures of unemployment, communication problems, whether a member has a chronic illness, a recent experience of divorce, death, or desertion, and so forth). For example, Schmidt's review of the literature back in the 1970s found that high household stress correlated with a steady increase in streptococcal illness, and a rise in antibodies. High stress also was related to increased risk of stroke, angina or chest pain, and concern with one's own and one's children's somatic (physical) symptoms. Pregnant women with high stress and low levels of social support had a pregnancy complication rate of over 90 percent (Schmidt 1978).

Schmidt also found that when one's spouse has a disease, one's own somatic symptoms were likely to increase. Similarly, certain chronic diseases appeared in both spouses at significantly higher rates than might be explained by chance. There were links between spousal deaths from a number of causes, including tuberculosis, hypertensive heart disease, influenza, and pneumonia.

Likewise, divorce has long been known to be unhealthy because of the stress it entails in certain contexts. Where divorce is not condoned, or when only one party seeks it, stress is heightened. Of course, divorce does not have such an impact in cultures where it is non-problematic or where other arrangements, such as serial monogamy, are the norm.

Stress in Cultural Context

Responses to Events Are Culturally Relative

Many people believe that contemporary humans bear a much higher stress load than our foraging ancestors did. However, our ancestors are long gone; we have no direct way to test this hypothesis with, for example, comparisons of salivary cortisol levels or aggregate allostatic load indexes. Even if we did, such comparisons would not tell the whole story.

This is because peoples' responses to events are culturally relative. The events and issues that create high and chronic stress for me as a member of the professorial occupational subculture of academics frequently are considered idiotic by members of other occupational subcultures (including some members of my own extended family, which adds to the stress). Conversely, not having an up-to-date wardrobe is no problem for me while it may be for some of my students, just as not having a particular tattoo might be problematic for someone whose tribe expects it but not for someone whose social network does not. Many expectations only are stressful when unmet in particular cultural contexts.

One time in Jamaica, where I conducted my initial dissertation field-work, I was doing some laundry. In the process of hanging out my clothes to dry, I walked across the little river that my old wash water had made in the dirt. You might have thought that I had cut myself with a straight razor considering the consternation this caused for Grandma and Auntie May, who were nearby at the time. They admonished me fretfully, concerned about the implications this might have for my health due to what might rise from that water, where it might go, and what it might do. Walking over wash water would be very stressful for a person brought up in their tradition, due to the understandings referenced. Conversely, for me, at least until I learned to fear these things, crossing a wash water stream was no big deal.

Another example illustrating the relativity of stress concerns our physical expectations. In cultures that valorize youth, people want to look young. At the least, they dye their hair; those with more resources and further commitment to the ideal get facial and other surgeries. Those who are most committed to the ideal do whatever is within their means to

Figure 10.3 Toposa elder in South Sudan. *Photo by Steve Evans, 2011. CC-BY-2.0.*

achieve and maintain a young-looking exterior. They feel ashamed and anxious when gray hair roots or wrinkles show.

In other cultures, youthful looks are equated with immaturity and a lack of achievement in life. In such cultures, the aged instead are held in high accord (see Figure 10.3). People with gray hair are respected. If you don't look old and worn out when the expected time for that comes, people may talk. They may assume, as they do among the Hua of Papua New Guinea when older people are too vital, that you have not been contributing to the community at the expected level—which, if you had been, would have led to proper aging (see Meigs 1997). In such a context, an older person naturally endowed with a youthful face and body would feel worried by and about his or her appearance and others' reactions to it. In such a culture, youthful-looking elders might eat less overall in an attempt to look frail; certainly they would eat fewer foods known to vitalize. They might expose their skin to extra sun or abuse their teeth to simulate the wear and tear of the average aging process.

I will give just one more example here, and that concerns teen heterosexual activity. In communities where virginity is sacred, the teenager who has had sexual intercourse, even if only one time or by force, may live in fear of being found out. Or, in contrast, take a community for which the youthful initiation of male-female sexual interaction is the norm. Here, a boy or girl who either has not yet managed a heterosexual partnership or a young person who is disinterested might feel worried and anxious. Concurrently, a homosexually oriented teenager living in a culture to which adolescent homosexual activity is unremarkable (such as in traditional Polynesia) would have no sexuality-related worries. A narrowly heterosexual male youth living where male homosexual relations are prioritized and heterosexual ones thought highly polluting (such as in some Malanesian groups) may, on the contrary, feel very stressed indeed.

Resilience

Just as sources of stress may differ, so do sources of **resilience**—strength in the face of otherwise stressful situations. Resiliency behaviors create functional competence for those who otherwise might pack it in under pressure. They include spiritual practices, such as magical acts attempted or undertaken in an effort to control the uncontrollable; humor; use of physical outlets such as sports activities; and reliance on networks of friends and relations for social support.

The size or extent of a social support network is not in itself a source of resilience. Having many friends or relatives interfering in your affairs can in fact be stressful in its own right, as anyone who has lived in a small village or with a large extended family knows (and which undermines the assumption that our ancestors were by definition less stressed than we are today). Resilience comes instead in the buffer resources provide by that network. These may include information, childcare, financial support, or assistance in requesting supernatural help, for instance through sacrificial rites or prayer meetings.

Because things that are stressful in one culture are not necessarily stressful in another culture, and sources for resilience differ somewhat, too, it is difficult to contemplate the development of a globally valid measure for stressful events. Because of this, and because biological measures of allostatic load are unavailable for our ancestors, current comparisons between the stress of contemporary life and the purported stress people experienced in the old days remain purely conjectural.

Stress Inequities: The Case of Race

We cannot validly weigh the past against the present in this case, but we can broadly compare the experiences of populations living side-by-side today. We can compare, for example, the stress experiences and allostatic loads of men and women in a given society, or of members of different religions. There are many directions we could go in setting up such comparisons, but the direction in which I would like to turn now is back toward social structure—and then forward to culture, which provides ideological support for the structure that a given society may have. I want to highlight social structure because, as the last chapter indicated, some social positions entail more long-term stress than other social positions. To bring culture directly into the picture, the positions I want us first to think about here are racialized ones.

To begin, let us take, as an example, the case of Kim Anderson, described in the *Unnatural Causes* documentary, "When the Bough Breaks" (California Newsreel 2008c). Ms. Anderson is a successful lawyer, having graduated from Columbia Law School in 1984. She was well-paid and in good health in 1990—the same year she had her first child. Also and by the way, Ms. Anderson is African American.

One might expect excellent outcomes for her and the baby. True, infant mortality rates are twice as high for black people in America as for white ones (MacDorman and Mathews 2008). At the same time, many African Americans also are poor (see Barr 2008, 134–168). We know from Chapter 9 that occupying a lower class position in society—having a disfavored location in the social structure—is tightly linked with poor health outcomes due to inequities in access to healthcare and to unhealthful living and working conditions.

Yet, it turns out that even babies born to highly educated African American mothers like Ms. Anderson are at a higher risk for premature death than is true for the population as a whole. Infant mortality among these moms is two and one-half times that of equivalently educated white American mothers.

If the difference were racially-generated (a hypothesis that only works if one presumes the existence of human races, but see Chapter 5) we would expect to see high infant mortality rates in all populations with African heritage. Yet research shows that this is not the case. One study, for instance, compared the newborns of African American women, African immigrants to the United States, and U.S.-born white women. While the African immigrants' babies and the whites' babies were generally the same, the babies of the African American mothers were different. They were about one-half pound lighter—and for a baby, one half pound can make a significant difference. Furthermore, after one generation of living in America, outcomes became significantly less favorable for the African immigrants as well. So low birth weight and high infant mortality rates among blacks are not racial or 'natural' outcomes; rather, something environmental—something cultural—seems to be at work (see also David and Collins 1997).

Kim Anderson is, as noted, African American. In regard to her life-style, as she tells it, "People would think I'm living the American Dream: a lawyer with two cars, two and a half kids, the dog, the porch, a good husband, great family. I've always been lucky to have good health. Always ate well. Exercised. Never smoked." During her pregnancy, she says, "I did all the right things. They told me to take vitamins; I took vitamins. They told me to walk. They told me to eat vegetables. They told me not to drink." Yet more than two and one half months before her due date, Ms. Anderson went into labor. When her baby was born, she said, "I heard her cry, I said, 'Thank, God.' But she was so small. [You could] hold her in the palm of your hand" (California Newsreel 2008c).

Why did things turn out that way? Stress. Remember, Ms. Anderson is black, just like the other African American mothers whose pregnancies did not go full-term, and just like the daughters of African immigrants who went on to try to build families in America. Thereby, every day of her life, she has a chance to experience racist treatment.

Although many believe that racism is no longer a problem in the United States, even now African Americans are more likely than others to experience prejudicial treatment, for instance in being charged more for equivalent automobiles, being denied mortgages, and receiving fewer job interviews. In a study done in 2004 in Milwaukee, Wisconsin, black and white men (actors) were sent to apply for 350 entry level jobs, all with exactly the same resume. The black men fared worse. Moreover, black men with no criminal records were less likely to be offered jobs than white men with a felony conviction (David Williams, in California Newsreel 2008c).

Work from 2010 in a similar vein demonstrated that Craigslist ads for iPods shown being held by black hands received 13 percent fewer responses and 17 percent fewer offers than those in which the iPods were in white hands. The dollar value of offers received for the iPods displayed in black hands was 2 to 4 percent lower. Buyers corresponding with black-handed sellers also displayed lower levels of trust: they were less likely to share their names or to accept a delivery at their mailing addresses (Doleac and Stein 2010). A lifetime of such discrimination—of living in a racist culture—surely can add up.

At least in the United States (and in many other places, too), Ms. Anderson and others who belong to minority groups live chronically stressful lives because the world they live in is racialized. So, while about half of all white women report that they never think about their race, about half of all black women think about their race at least once a day; about 20 percent think about it all the time (Camara Phyllis Jones, in California Newsreel 2008c).

When Ms. Anderson walks into a shop, or a bank, or just down the street, the first thing many people notice about her is that she is black—and they adjust their actions toward her according to stereotypes held regarding the 'black race.' In the documentary cited, Ms. Anderson recounts a time when she walked into a shop, thinking to buy some jeans. The clerk started following her around, as if she was going to steal something.

Now maybe the clerk wanted to advise Ms. Anderson of a sale, or keep her away from wet paint. Nevertheless, based on her past experience of racist treatment (which, as we saw above, research does confirm is quite real), Ms. Anderson felt otherwise—in the flesh, quite literally. Describing her response in the face of racism, she made physical reference to the anxiety it provoked: "your stomach just gets like so tight. You can feel it almost moving through your body; almost you can feel it going into your bloodstream" (California Newsreel 2008c).

Racism can, in fact, result in heightened stress in those who experience it. We already have established that chronic stress writes itself into the body, for example altering heart function and insulin-related processes. The bodies of people under chronic stress are therefore more prone to diseases such as hypertension and diabetes. They also are more prone to premature labor, which contributes greatly to infant mortality rates. This may be because some of the hormones generated in stress also are involved in triggering the start of labor. If the body begins pregnancy with an already high load of stress hormones, it may take less time to reach the tipping point at which contractions begin (California Newsreel 2008c).

Premature births and low birth weights represent just some of the many health problems more prevalent in African Americans. One report (Drexler 2007) noted that more than 100 studies had documented the link between racism and health status. Of course not all focus on African American health, but that other minority group members experience similar outcomes reinforces the contention that the stress of racism, and not race itself, is at work. For example, one study showed that after the September 11 attacks on the United States, pregnant women with Arabic names had a sudden surge in low birthweight deliveries.

Mounting evidence shows that health problems such as hypertension, diabetes, and infant mortality—long known to occur at higher rates among black people than among white people in the United States—are in many ways the result of our racist culture. In an interesting twist, these conditions may be exacerbated by the cultural myth that racism is no longer a problem: this myth intensifies the stress that accompanies racist treatment because it raises expectations for nonracist treatment from the start, and implies that racism will not be seen by others as having been real (see Wyatt, Williams et al. 2003; see also Mays, Cochran et al. 2007; California Newsreel 2008b; California Newsreel 2008c)

Individual Versus Population Risk: Stereotypes Versus Generalizations

Importantly, all the problems we are talking about are population-level problems. That does not mean that they do not affect individuals; they surely do. In fact, I just used a person-focused case study to give life to the issues under discussion. However, the figures provided and the extrapolations regarding risk are not about individuals. They are about *groups*.

It is logically problematic to go from a population-level statistic to an individual. Just because an individual is a member of a given population does not mean that he or she will have the same outcome that is seen at the population level. This is not only because that individual belongs to lots of other groups, too, which muddies the relevance of the statistics. It is because logically speaking—mathematically speaking—individual risk and population-level risk are two different things.

To make this a bit clearer, we might take, for instance, population-level statistics regarding a college course on biocultural diversity. The instructor might tell students that the class as a whole, or the students on average, scored 75 percent (earned a C) on a given test. The average score holds for the group, but it tells single students very little about their individual scores. A student's score may be much worse—or, for those who have gotten this far in the book, more likely much better. A student can compare his or her individual score to the average—but that student cannot guess his or her own score on the basis of that average alone. Likewise, you may note that members of a group you belong to have a 5 percent chance of developing hypertension, but there is no way to jump from that figure to predicting your own odds for the disease.

When someone takes a population level statistic and applies it across the board to any individual belonging to that population, that someone is stereotyping. As Geri-Ann Galanti has noted, a 'stereotype' is an end point. A 'generalization,' however, can be seen as a starting point (2008). Knowing what we know about population-level statistics—that is, based on generalizations we make from the statistics we have—we can move forward and implement public health programs that can raise the health status of entire populations.

Cultural Consonance

The term 'consonance' is somewhat like concordance: it means that things properly align. **Cultural consonance**, a construct developed by William Dressler (see Dressler and Bindon 2000), is the degree to which one's lifestyle fits with the lifestyle that one's culture recommends and that one thereby aspires to. When the fit is bad—when we have not achieved the things that we had in mind to achieve by a certain point, or the things that our culture has taught us to want to achieve (such as owning a particular kind of car, having earned a college degree, or being married with children, a dog, and a white picket fence)—cultural consonance is deemed low. We therefore feel stressed. More concretely, health outcomes suffer.

Dressler and colleagues have examined the link between health outcomes and cultural consonance in several cultures (including in Brazil, the Caribbean, and Alabama) via clinical biological measures, symptom self-reports, and ethnographic methods. This body of work has demonstrated the link between high aspirations, low levels of the resources necessary to achieve them, and poor health outcomes such as high blood pressure. It also has shown how community resources for resilience can serve as a buffer.

The concept of cultural consonance subsumes those of role incongruity and status inconsistency, which are in effect other ways to talk about somewhat the same thing. Indeed, Dressler used such terms prior to inventing his cultural consonance measure. Sometimes these terms are used interchangeably. Nevertheless, I will define them discretely here.

Role incongruity occurs when one is not living up to the expectations entailed in the role one has been placed in or has elected to take on. For instance, a student may experience role incongruity if he or she has some idea of what a good student does and achieves but cannot seem to get these things done or get his or her grades up. The student feels a gap between the role as he or she wants to play it and as he or she really is playing it out. Another example might be when someone finds him- or herself a father or mother and cannot quite live up to his or her own internalized cultural expectations of what makes a father or mother 'good.' When there is a gap, there is role incongruity, and where there is role incongruity, there is stress.

Status inconsistency is similar but directly relates to how others view someone. Status inconsistency occurs, for instance, when someone belongs to a profession that was valorized when he or she entered it but is no longer held in such high esteem; doctors sometimes report feeling this way. Another example is the refugee who in his or her home country had a very high status job or position, but who in the new nation ends up doing a low status job and is treated as a low status person by members of the host nation. Members of minority groups who find themselves discriminated against also can experience status inconsistency.

Status inconsistency can be felt, too, when one has the outward trappings of success but not the real basis for it, as when a person who seems to be living well is, in reality, deeply in debt, or when someone with a good job actually does not have the education necessary for it. While some might class this kind of situation as role incongruity, others do not, because in this situation a person can keep up appearances—albeit with difficulty. With role incongruity, a person is an outright failure in fulfilling the expectations of a given role.

If a person is either under pressure to keep up appearances or accustomed to respect and now gets none, he or she feels bad. Feeling bad over a long period of time can trigger disease processes, including sometimes mental breakdowns.

Culture-Bound Syndromes to Channel Stress

Our examination of the impact of chronic stress makes it clear that the physical body (biology) and the mind, which interprets events (culture), are intimately connected. Beyond stress, another way culture gets into the body to create and sustain biocultural diversity is via what have been termed **culture-bound syndromes**.

What Are Culture-Bound Syndromes?

Culture-bound syndromes are culturally unique, culturally named conditions. They bring together, as syndromes, a variety of symptoms, most of which would otherwise (in the mainstream United States) be written off as 'stress' and many of which may not have any biomedically related explanation for concordant or concurrent (syndromatic) expression.

While culture-bound syndromes initially were studied as unique and taken simply as indicators of cross-cultural diversity among humans, more recent work highlights their universal features. For instance, many entail what is termed **somatization**—the projection of mental attitudes or concerns onto the body so that they are expressed as physical symptoms. As a simple example, think of the school child who gets a stomachache every time that his or her teacher schedules a test. Anxiety, distaste, annoyance and the like actually can be felt in the stomach, head, and neck; we do experience pain or become tense in these areas (we somatize) when such feelings cannot be otherwise discharged.

In addition to their high likelihood of featuring somatization, culture-bound syndromes often provide an outlet or channel for stress brought about by role incongruity. In doing this, cultures provide stressed individuals with safety valves of a sort. As McElroy and Townsend note, "Rather than becoming and remaining labeled as mentally ill, the person is considered to be a victim of witchcraft, soul loss, severe shock, vengeful ghosts or other forces. After recovery, little stigma is attached" (2004, 280).

Traditional Examples

For example, in many Spanish-speaking societies, an individual who experiences a sudden fright may have his or her soul scared or taken away; this person becomes *asustado*: he or she is a victim of *susto*. Susto's symptoms include loss of appetite and weight, apathy, withdrawal, and depression; nightmares, swollen feet, diarrhea, general pain, and headaches also have been reported. While asustados experience higher rates of organic disease than those who do not suffer from susto, this is an outcome of their vulnerability; there is no pathogen or lesion causing the primary condition. Rather, it seems to be brought on by stress—the stress caused by role incongruity.

Many susto sufferers experience some kind of crisis of inadequacy just prior to the susto attack. They may have been socially sanctioned through being laughed at or shamed; their household partners may have been cheating or using household savings for personal gain; they may have experienced something that made them feel unable to carry out ascribed social roles. The discrepancy or gap between their actual or socially viewed performance and their own expectations for a job well done caused them to experience excess stress (Rubel, O'Nell et al. 1984).

Modern Culture-Bound Syndromes

There are culture-bound syndromes in so-called modern cultures, too. Menopause provides one example—and here the term is used not to mean the point in time of the last menstrual cycle, which is its clinical meaning, but in its popular sense, to refer to a syndrome including mood, body temperature, and sleep disturbances affecting older women when menstruation is ceasing. Menopause in this sense, which does not happen everywhere, can be understood as a culturally acceptable way of diffusing or grappling with the stress of the gap between the cultural equation of femininity with childbearing and youthful beauty, on the one hand, and the impact of aging on the other. The menopausal woman in this cultural context is not the woman she wishes to be.

If this interpretation has any truth to it, in cultures where postmenopausal women are culturally valued contributors, or where their social status increases postmenopausally, the syndrome should not be seen. Indeed, although the average age of 50 is remarkably universal for onset of menopause, a survey of thirty societies conducted before the recent and potentially confounding surge of globalization revealed that only in two societies besides the United States was the event seen as a major physical and emotional loss (Bart 1969). The accompanying symptoms of distress found in the United States were not even noted in other societies such as India (Flint 1975).

McElroy and Townsend also bring up so-called road rage as a Western example in their review of culture-bound syndromes. Road rage entails the sudden onset of extremely aggressive behavior among (mostly male) drivers. Its occurrence has been documented in the United Kingdom, Holland, and the United States (2004, 286). It may be an expression of the stress engendered by the gap between being treated or received on the road as one wishes to or believes that one should be treated or received, or it can be an expression of such feelings of status inconsistency in other areas of life, or of role incongruity. It may be similar to syndromes such as *amok* or *amuk*, a temporary kind of insanity which occurs in New Guinea, Indonesia, and Malaysia, mostly among young men. A victim of amok goes suddenly frenzied and runs about, trying to injure or kill anyone in his path. The English expression 'running amuck' stems from colonialists' knowledge of this condition.

The late twentieth century upsurge in anorexia nervosa among white, middle-class teenage girls in the United States also has been analyzed as a culture-bound syndrome. Anorexia nervosa is an extreme form of self-starvation linked with an intense fear of being fat. On the surface, it can be linked to an overabundance of cultural messages regarding the value of a slender body, particularly for women in the white middle class (in other groups, being plump can be taken as a sign of vitality and good health). However, at a deeper level, anorexia nervosa also involves rebellion against the social relationships represented in food-sharing. It has been analyzed as an enabling condition of teenage girls who, particularly in white middle-class families with controlling parents, may be having difficulty individuating or otherwise feeling empowered over their own destinies.

The Biocultural Nature of Culture-Bound Syndromes: *Pibloktoq*

While conditions such as road rage, amok, and susto can be understood in purely cultural terms, with culture playing the key role in triggering the stress responses that they channel, anorexia nervosa and 'menopause' both demonstrate more complexity in terms of how the biology figures in. One culture-bound syndrome exemplifying the complex synergy that can be in place between culture and biology is *pibloktoq* or "Arctic Hysteria" (McElroy and Townsend 2004, 280–283), in which affected individuals suddenly begin screaming and making animal-like noises, tearing off their clothes, running away, throwing things, and possibly rolling in the snow or jumping in icy water. Prior to an attack, a person may exhibit irritability, confusion, or depression. Afterward, she falls deeply asleep. When she awakes, she remembers nothing, and functions as normal. I use the pronoun 'she' here because most victims are women. Also of note is that pibloktoq occurs mostly in winter.

Pibloktoq is treated by Arctic peoples as a physical malady. Biologically oriented researchers have suggested that pibloktoq actually could stem from a calcium deficiency or vitamin A overdose. Both of these things are most likely in the winter, when the majority of pibloktoq attacks have occurred. For one thing, the light-dark cycle in winter seems to be linked with biological 'desynchronization,' so that more calcium is lost through the urine than normal; for another, the shortness of food supplies in winter can lead people to rely more on organ meats high in vitamin A, like liver (the fat of arctic animals and fish also are high in this vitamin).

Other studies have shown that prior to an attack, victims very frequently experience emotional conflict, shame, and anxiety, often in regard to an inability to fulfill household and community role expectations. Such stress may leave these people in particular more vulnerable to the effects of excess vitamin A or calcium depletion, both artifacts of cultural adaptations necessary for Arctic living.

Culturally Interpreted Disease States

Until this point, my focus has been on syndromes expressed in behavioral and mental states, even when somatization is identified. However, there are syndromes with more narrowly physiological signs and symptoms. The convenient distinction between body and behavior is, of course, ethnocentric. It also flies in the face of this chapter's contention that body and mind are interrelated. That said, it still may be helpful here to explore some ailments that are (or seem to us to be) more body-bound—ailments in which symptoms themselves are overtly physical rather than attitudinal. These include conditions brought on by cultural practices, such as late nights spent staring at a computer screen or long spans of time spent sitting in front of one. They also include disease states or problems found around the world in one form or another, but in each place with a distinctly cultural twist.

Empacho provides a clear-cut example of this kind of culturally relative physical malady. Empacho, which is most common among people of Latin American cultural origin, entails a kind of stomach distress—the kind that happens when food gets stuck to the inside of one's stomach or intestine. The food can become attached though suction, much as the soft bread of a peanut-butter sandwich can adhere to the roof of the mouth. In empacho, the food stays stuck, however, sometimes having trapped other substances between it and the wall of the belly or wherever it has lodged. It stays stuck long enough to start to rot. Whatever is trapped starts to rot, too. The whole conglomerate begins to generate gas and swelling, leading to a lack of appetite that can be accompanied by headache, diarrhea, and vomiting.

The sensory feelings that I am describing can be felt by anybody, anywhere in the world. We all get stomach upset sometimes. The feelings reflect universally measurable biochemical processes and concomitant signals and impulses traveling through the nervous system. Yet, depending on culture, people may interpret the symptoms entailed quite differently. Those who grew up in a Latin American culture will have been provided, since childhood, with the cultural construct of empacho. Knowledge of the condition provides a person with a mental tool through which to interpret the signals of abdominal fullness and so on—and with ideas on

Figure 10.4 Beecham's Pills, a laxative treatment made from aloes, ginger, soap, and several minor ingredients, came onto the market in England in about 1842. They continued to be manufactured until 1998.

what to do to treat them. With empacho, the goal is to loosen the stuck or impacted food so that the stomach is emptied again and ready for regular use. For instance, in Guatemala, people report resorting to stomach massages, over-the-counter remedies containing sodium bicarbonate (an alkali and so an antacid), lemon juice, purgatives, and herbal teas, some of which also are purgatives. Another option is to take some olive or cod liver oil or to ingest something ashy, such as a burnt tortilla (Weller, Ruebush II, Klein 1992). Other cultures have their own culturally influenced responses (see Figure 10.4).

Those without the empacho tradition have other ways to interpret and react to the sensations of a stomach upset. A mainstream white middle- or upper-middle-class American might interpret those same signals in terms of an acid imbalance. While he or she might not choose to have a stomach massage or to swallow ash, sodium bicarbonate or something similar likely will figure in his or her remedy scheme (bicarbonate is an ingredient in Alka-Seltzer).

Note here that there is some convergence in terms of medicinal strategies. Sometimes similarities stem from the diffusion of cultural practices: ideas do spread. Other times, people discover, independently, what works, and alkaline substances, including ash, do neutralize stomach acid around the world. However, diverse material or environmental options (what kinds of treatments actually are available) or varied views regarding an ailment's cause can lead to very different forms of treatment across cultures. If an illness has, say, a supernatural cause in one culture and not in another, the two cultures' remedies might be quite different or their administration will be differently approached. There is also a chance that one culture may subscribe to 'homeopathic' views, in which symptoms are treated by intensifying them, with the idea of helping the body to do what it intends to do anyhow to get rid of the problem (fighting fire with fire); another might take an 'allopathic' standpoint, in which the symptom is countermanded (fighting fire with water). So, for instance, someone with a fever might be wrapped up in a warm blanket or otherwise helped to further heat the body in one culture, while in another he or she might be given a cooling bath.

Beyond differences in treatment plans, or in the interpretation of symptoms from population to population, the very experience of stomach upset is bioculturally diverse. I did say that at the biological level the signals would be the same worldwide; that is true. However, because at the interpretive level a stomach upset might carry such a different meaning from one culture to the next, it actually can be experienced very differently. In a sense, when comparing the stomach upsets of, say the Guatemalan peasant and the Wisconsin-born Anglo American, we are comparing apples and oranges, or tortillas and kaiser rolls.

Yet another experience will emerge for someone whose culture provides not a naturalistic (for example, food or stomach-acid based) explanation but rather a personalistic one, based on how one's relations with others are faring. In such a context, a stomachache could signify social rejection, homesickness, an angry neighbor, or any number of problems within one's social network. Treatment plans will focus not only on fixing the stomach, but also on—and sometimes instead of—fixing one's social situation. Treatment will be not individualistic but rather community based.

Ending this section with such an example helps drive home an overarching theme of this book: folk traditions notwithstanding, it is unhelpful, at best, to try to explain scientifically any human experience as either wholly biologically grounded or completely cultural. Our own language sets us up to make this mistake. We have no unified word for the mind-body, and so we do tend to see ourselves dualistically even when trying not to. Yet, because biology and culture (or nature and nurture, or body and mind) are actually a unity, only a unified, integrated approach can provide real answers, and real understanding.

Placebo Effect

The Meaning Response

At this point, we must consider the potential objection that much of what passes as healing is not actually 'real' medicine. A community ritual held for a person with a gastrointestinal ailment is not what 'really' heals it. Rather, the educated reader may say, something used in the ritual had clinical efficacy; the ritual itself made no difference; or the illness just ran its course and everyone simply misattributed its disappearance to the concurrently timed ritual. A common cold, for instance, eventually will get better on its own with or without a healing ceremony. Another explanation for why people believe in ritual healing is that they are simply taken in by the placebo effect.

A **placebo effect** is a measurable, observable, or felt improvement in health that is not directly attributable to biomedical treatment. So a **placebo** is an inert substance or an act that is not part of the biomedical tool kit, such as the prayerful laying on of hands or other forms of faith healing; the placebo is the act or item that's not doing anything. The placebo effect is generally ascribed to self healing attributable to knowledge or belief—to the cultural meaning—that a practice or process carries. Daniel Moerman (2002) suggests, on this basis, and because an effect cannot logically be named after or caused by something that by definition has no effect (that is, a placebo), that instead we should label this kind of healing 'the **Meaning Response**.'

However labeled, this response or effect is well utilized in many cultures. It is not unknown in biomedicine, where cures or improvements often are created with inert substances given to patients who believe them to be medication. For example, as much as 50 percent of analgesic pain relief is due to a placebo response (Watkins 1996). Even a prescription itself can have a placebo effect; as Norman Cousins noted years ago, leaving the doctor's office with one in hand is a major part of the expected healing ritual (1981).

Although we do not know exactly how most placebos make a difference, we can be certain that what works as a placebo is culturally relative. A placebo's success relies on the patient's socialization, beliefs, and trust in the healer. As Jamaicans say, belief can cure, and belief can kill—although placebos with negative effects are more correctly called 'nocebos.'

Fear Kills

Some of the earliest work regarding the negative power of belief was undertaken by Walter Cannon, also one of the first to help define the biochemical pathways for the stress response. Cannon was interested in explaining 'unexplained' deaths among those who believed they had been hexed. After reviewing a number of cases in which Australian Aboriginal and Maori as well as native South American people had died after learning that someone had cursed them, Cannon realized that their deaths had to do with the stress response.

While to cultural insiders the hex was the root of the problem, Cannon explained this kind of death, from the biomedical perspective, as due to the traumatic stress brought on by intense and unabated fear: he argued that when epinephrine (then called adrenaline) runs high, blood vessels constrict, lowering the amount of circulating blood; vital organs are therefore not well-enough oxygenated to support the body. On this, Cannon actually was more or less correct, although newer work on allostasis, such as we already have discussed, has broadened our understanding of what is really going on. In any case, as Cannon showed, if we add to low blood levels and so on the lack of sleep due to worry, no intake of food due to belief in one's sure demise, and social rejection or 'social death' of the hexed person (for whom relatives already might be digging a grave and singing funeral dirges), death is in fact a certain result (Cannon 1942).

Meaning Matters

Proximate, biomedical causality such as Cannon described notwithstanding, it turns out that, ultimately, a person's culturally directed interpretations of his or her situation are what underwrite biomedically inexplicable—but physiologically measureable—healing (or pain reduction, or death). To make this clear, let me recount a classic example (Blackwell, Bloomfield et al. 1972, as described in Moerman 2002, 47–51).

About forty years ago, professors at a medical school undertook a study with fifty-seven second-year medical students who had come to class to hear a lecture. At the start of the lecture they were told that scientists had discovered two new drugs, a stimulant and a sedative, and that they would be part of an experiment to test the new drugs. The students were randomly divided into four roughly equal sized groups and then given pills—which were inert. The groups respectively received either one pink pill, two pink pills, one bluish pill, or two bluish pills. The students were told only that the pills were sedatives and stimulants; they were not told which pill was which.

Before taking the pills, the students took each others' blood pressure and completed a self-report mood survey. After taking the pills, they listened to a one-hour lecture, as medical students commonly do. After the lecture, they again took their blood pressures and reported how they were feeling.

What did the findings indicate? They showed that one quarter (26 percent) of the group that took one pink pill felt less alert after the lecture, while two-thirds (66 percent) of the group that took one blue pill felt so. The pink pills seemed to inoculate or protect the students against lecture-linked boredom much better than the blue pills did. Furthermore, students who got two pink pills were more protected than those who got one pink one, and those who got two blues were less protected than those who got one. So doubling the dose seemed to intensify the effect. Remember, however, that the pills were inert; they did not have any kind of pharmaceutically driven impact on the body. Everything that happened to the students, then, stemmed from the meaning response.

As members of mainstream American culture, the students had learned to associate blue with coldness, slowness, sluggishness, night time, and sleep; they had learned to associate pink, as blue's opposite, with warmth, heat, and activity. The students associated the colors of the pills they had taken with the states of slowness and quickness. Note that in other cultures the meaning of these colors can be quite different. Nevertheless, given the students' cultural assumptions, the pills' color-related meanings—through the mind and through the brain—affected the body: the pills worked—and they worked as the students expected them to. The connotations of the numbers of pills also played a role. We live in a culture where more is better, so again, apparently naturally and without effort, but in a way that really was very cultural, the students assumed that if they had double the dose, they would have double the effect. Of course, being second year medical students interested in helping people with necessary pharmaceuticals, they also assumed that pills must have some kind of an effect to begin with, which in itself supported their experiences of the pills' impact.

Recent research has shown that even knowledge about price can support healing. In a U.S. based placebo pill pain reduction study involving electric shocks, subjects given the placebo but told that it was rather expensive ($2.50 per pill) fared better than subjects given the placebo but told it was only 10¢ per pill (Waber, Shiv et al. 2008). The adage, "You get what you pay for," or the assumption that items that cost more are better may have been at work.

Similarly, studies regarding the role of the built environment in facilitating or impeding healing in hospital settings have supported the use of features such as garden views for patient rooms. For example, Roger

Ulrich (1984) was studying people's outcomes for gallbladder surgery and did a comparison. He followed patients who were put in hospital rooms either looking over a grove of trees or looking out at a brick wall. These were comparable patients, who had had comparable surgeries. Those who were in the rooms that looked out on the grove of trees had shorter stays and used less pain medication than those who were looking out at the brick wall. This study has been repeated over and over again, in diverse forms and ways. So for example, researchers have found that patients who stay in sunny rooms use 22 percent less pain medication (Ulrich 2006).

Their improvement, the improvement of others like them, and the improvement of others in the care of nonbiomedical healers (such as curanderas, shamans, and priests) seems to be the result of the treatment's meaning, or of some meaning attributed to the context in which the treatment is given. That is, a pill's color, name, shape, or even the dress and demeanor of the person dispensing it, the extra attention entailed in its provision, or what the patient is told to expect can underwrite a degree of healing, or pain reduction, or the like—sometimes enough so that researchers comparatively testing a new drug or procedure against a placebo cannot legitimately or legally claim the drug or procedure to be in itself a significantly better source of healing.

Leveraging Knowledge

As Moerman (2002) notes, rather than dismissing the placebo effect as 'unreal,' we should find out what specifically in or about the treatment supports self-healing. If we knew more about the connotations that have helped make placebos so effective in various cultural contexts, including those that support ritual nonbiomedical healing—if we knew more about the mind-body, culture-biology connection—we could begin to leverage the placebo effect to optimize the meaning response and improve patient outcomes. Similarly, if we knew more about allostasis and the stress response, we would be better equipped to stem its onslaught.

Meaning—culturally constructed meaning—acts directly on the body, triggering a helpful or a pathogenic response. Yet, while belief can cure and belief can kill by provoking a meaning response, it does not always. We must acknowledge the limits of belief. Some diseases are much easier to cure through a meaning response than others. On certain conditions, it works fine, but on others—cancer, HIV infection, malaria—its impact is palliative at best. Humans are wonderful and amazing beings, but nobody lives forever, and the body, even with culture's help, cannot do everything by or for itself.

In any case, in the meaning response, and in chronic stress, cultural meanings are embodied. Just what diverse populations embody will differ depending upon the cultural context in which each population has been raised. We saw, for instance, how events may be perceived entirely differently in distinct cultural contexts or for different social groups, leading to high stress—and all that that entails—in one setting or group, but not elsewhere. We even saw how racist ideas in fact underwrite certain measurable differences between so-called races. Culture's power in shaping biology, or in triggering evolved biological responses in ways that vary from population to population, should not be underestimated.

Dynamics of Human Biocultural Diversity: A Unified Approach by Elisa J. Sobo, 207–232.

Gender Diversity

This chapter prepares you to:

- Explain how culture is embodied in its practice
- Demonstrate the difference between sex and gender with reference to gender diversity across cultures
- Explain the conceptual basis, power source, and cultural functions of crossover or anomalous gender roles
- Characterize and apply the Rites of Passage framework to explain how social transitions are accomplished and to demonstrate how expectations for manhood and womanhood are culturally constructed

Introductory anthropology classes often present culture as a set of rules that people follow—a design for living—or as the superorganic 'thing' that emerges when we follow those rules. In advanced classes, however, and beyond, many argue that culture actually does not exist at all except for when we put it into practice. It comes into being in that process; it is a processual, event-bound, emergent creation. It emerges through actions we undertake with the use of our bodies.

Thus, a deeper look at the 'practice approach' is warranted for this text. After demonstrating the importance of practice for our inquiry into biocultural diversity in general, we explore it in relation to gender in particular.

I should caution up front that our focus in this chapter is on biocultural diversity as demonstrated, and in some cases created, through gender practices. We will not have time to examine sex diversity, or diverse sexualities—although those are interesting and worthy topics. Rather, our focus will be squarely on how cultures (their members) create and enact gender—how they envision and practice masculinity and femininity (which, as readers will see, may have nothing to do with sex or sexuality in particular populations). We also will explore the way that rites of passage are used to help children make the transition from being children into adults who rightfully can lay claim to, and must bear the responsibilities of, full status as members of a particular gender category.

Why Study Culture in Practice?

I would not go so far as to say that culture only exists in the 'doing' of it, because each culture certainly also pre-exists a given act, driving people to do as they do. Still, it is helpful to think about culture as a process versus a thing because, among other benefits, this approach helps us to understand culture's stickiness or how it becomes such a habit. A practice's actions, repeated frequently enough, can come to seem so natural that we may not be able to conceive of any other way of doing said actions. Also, having one way of doing an action can predispose us to doing other actions in particular ways, too, as in a chain reaction. What is more, for better or worse, prominent action patterns can be highlighted by people who wish to argue in favor of reductionistic biological determinism as, for example, in regard to each gender's ability to throw a ball.

While we previously considered practices directly linked to notable health challenges, in this chapter we will think about perhaps less sensational but no less important ones, such as high heeled shoe wearing. In certain communities, wearing high heels assists people claiming femininity. High heels also affect the anatomy of the foot and leg if regularly walked in, in a way that can be drawn upon by those wishing to disparage women's walking and running abilities. What happens is that the Achilles tendons of long-term high-heel wearers become notably stiffer and thicker than those of people who wear flats. This apparently helps offset forced and measureable calf muscle foreshortening; it also leads to pain when standing or moving on flat feet—which keeps high-heel aficionados from doing much of that (*Economist* 2010b).

In addition to the way it helps illuminate the symbolic and practical effects of what people do, the culture-as-practice framework also allows insight into culture change. For it is in the process of 'doing culture' that adaptations can happen. Something that works really well in one context may cease to work well in a new context, and so people may begin to do it in a novel, better adapted way, introducing new variations into the biocultural mix.

Culture in Practice

No Body, No Culture

Culture cannot happen without our bodies. Our very existence as social beings is body-dependent. To make and maintain social relationships, we variously look at each other, touch each other, and talk to each other. Through talk—the specialized use of teeth, breath, tongue, lips, and larynx—we teach each other things: cultural things, like how to do a certain type of dance or how to eat or dress in a way that is culturally acceptable. Culture is embodied in a very literal way when we pay attention to the shape and the size of our body and try and sculpt it into the shape or size that our culture idealizes.

We even embody culture when we sleep: culture helps determine whom we sleep with, when, and where. In some cultures whole families sleep in the same room or immediate vicinity; in others, mothers sleep with young children and fathers sleep with other village men elsewhere; and in still others, where each nuclear family shares a home, only the conjugal duo shares sleeping quarters. Offspring may share a room due to lack of funds; but the cultural ideal in the latter case is, for children, a room of one's own.

Culture and Emotion

Emotional expression comprises another arena for culture's embodiment. For example, Japanese culture, like many others, favors emotional control as part of an overarching **ethos** (worldview; fundamental value set) that places the group's well being ahead of self-centered priorities. This is in contrast with mainstream America's ethos, which prioritizes self over society in many circumstances. Not surprisingly, then, when researchers led by Masaki Yuki showed Japanese students (at two Hokkaido universities in Japan) and American students (at Ohio State University in the United States) pictures of faces of people either hiding or expressing their emotions, the Americans were less successful at accurately gauging how each photographed person really felt. The Americans did not grow up needing to develop skill in deciphering hidden emotions and so they were worse at it than the Japanese (Yuki, Maddux et al. 2007; see also Nisbett and Masuda 2003).

The trick to accurately reading masked emotions, the researchers explain, hinges on looking at eyes rather than mouths. Eye muscles reflecting emotional states are hard to voluntarily control and so are more likely to give away our real emotions. Our lip and cheek muscles, however, are easy to override when we want to hide emotion. That is, due to the physiological and anatomical legacy of language, we can quite freely manipulate smiles, frowns, grimaces, and so forth if we want to. Japanese people are enculturated to look at the eyes, because they know mouths might be lying; they are thereby better able to discern hidden feelings than mainstream Americans who, lacking similar cultural training, naively focus on the mouth. This culturally-engendered capacity exists as an adaptive response to culturally-fostered emotional inexpressivity.

Variation in the way populations embody their emotions has implications not only for interpersonal relationships, which may in fact be enhanced by the ability to see what people really are feeling, but also for the unlikely arena of email and text message communication. For Japanese-born people and cultural groups with similar approaches to emotion, emoticons focus on the eyes; for others, emoticons emphasize the unreliable mouth (see Table 11.1).

Table 11.1 Comparable Emoticons

	Happy	Sad
Eye-focused	(^_^)	(;_;)
Mouth-focused	:)	:(

Culture and Cognition

I wish we had room for a full chapter on culture and emotion and one on culture and cognition. The variation we see in how people feel and think across cultures is eye-opening. It is the case that, by many indications, much human thought is universally determined (for example, binary contrasts such as up and down, right and left, good and bad are found across all cultures, and developing children in all populations seem primed to develop particular cognitive capacities in more or less the same order). However again, as with our more corporeal (physical) selves, different environmental conditions lead to populational diversity. This is not just in terms of content, as might be expected, but in the form of cognition as well.

Take, for example, the cognitive patterns found among U.S. youth today, bombarded as they are with diverse media and technology based stimuli. I teach online, and my surveys confirm that many students listen to and view online lectures while simultaneously texting and being texted by friends, checking and posting to social network sites, monitoring other particular websites, and so on. Some argue that current use of information technology actually strengthens the brain, much like going to the gym and lifting weights strengthens the body. People who use a lot of technology, or so the argument goes, are much better at, for example, multitasking than our electronically bereft great-great-grandmas and grandpas ever could have been.

Evidence, however, shows clearly that multitaskers are, in reality, often distracted, so that any net gains that might be had by doing many things at once are offset by reductions in how well each task is completed. The more experience a population has with multitasking, the less distracted they will be—but in the end, there is no free lunch, and overall performance for all multitaskers does suffer. In the words of Clifford Nass, whose research team was in fact "absolutely shocked" at their own counterintuitive findings: multitaskers are "terrible at ignoring irrelevant information; they're terrible at keeping information in their head nicely and neatly organized; and they're terrible at switching from one task to another." They (and we) are, however, good at believing the opposite (Dretzin 2010).

The collateral damage a population incurs from multitasking may be quite far-reaching. Habitually flitting from one thing to the next seems to be related to an inability to follow through in various areas and what some call a 'soundbite' mentality. Some researchers and clinicians believe that the dramatic increase we've seen in varieties of Attention Deficit Hyperactivity Disorder or ADHD in the United States stems from such constant overstimulation. This is certainly an area of research that deserves to be pursued, as does all work exploring the phenomenon of embodiment.

Culture and Comportment

Culture is perhaps most obviously embodied in how people move: this becomes clear when we interact with members of another culture and see people moving in different ways. Often how they move looks funny to us outsiders. Alternative ways of moving may look even funnier when we try them—at least before we gain mastery over movements not made before, such as sleeping in a hammock, moving quickly and quietly up a steep rocky path, walking on a moving hand-made river raft, or carrying a load atop the head.

Expected modes of movement—including not just how people sit, stand, or walk, but also how they gesture and even how they express emotions—differ across cultures, and they also differ between cultural subgroups. Take gestures, for instance. While a nod of the head means "yes" to most Americans, Greek and some other Mediterranean people traditionally have said "no" by a similarly vertical movement of the head (Gross 1992, 109). A mainstream American might beckon a person forward by wiggling the index finger out and in with the palm up and other fingers and thumb closed or bent in toward the palm. This could be insulting elsewhere, however; for instance, this gesture is used in the Philippines to call animals (Galanti 1991, 22).

We also train bodies about what kind of personal distance from others to attempt or expect. For instance, while many Americans, Canadians, and British people require relatively substantial space between themselves and other people (Gross 1992, 109, reports this at about four feet), Latin American, Japanese, and Arabs prefer relatively little (about two feet for Latin Americans, according to Gross).

More total differences in comportment also exist. Think for example of the Japanese geisha. Geisha—female entertainers in the classical tradition—are trained from an early age to hold their bodies (including their heads and eyes) in a certain way and to walk in a particular manner (see Figure 11.1). The geisha walk is of course to some degree constrained by the beautiful kimonos and shoes that a geisha must wear. Still, to say that is the same as saying that geisha are created and

Figure 11.1 Traditionally, geisha training entailed an apprenticeship. "Geishas" (n.d.). *George Grantham Bain Collection, Library of Congress, LCCN LC-B2- 5534-5A, LC-DIG-ggbain-32973.*

Figure 11.2 Geisha in uniform (n.d.)
*George Grantham Bain Collection,
Library of Congress, LCCN LC-B2-
5189-7, LC-DIG-ggbain-30458.*

constrained by culture: the kimono and so forth—let alone the form of walking—are cultural artifacts (see Figure 11.2).

It is true that geisha belong to a very special subculture in which practitioners undergo intensive physical training—not just for walking but for playing musical instruments, dancing, serving tea or other beverages, and so on. Yet even without specific training or openly acknowledged rules for comportment, from day one a person's method of movement is influenced by culture. It is influenced by cultural expectations for particular roles, genders, or professions. All movement is culturally influenced. Even the way that people interact with one another sexually differs across cultures. Sure, there are some universal similarities, but what is so interesting here are the things that are not universal—the diverse patterns expressed by people as members of particular cultures.

One way to test this (as my students do in live classes) is to play charades, in which people are instructed to act like members of particular cultural groups. If given, for instance, a card that says "cowboy," a player will know to bow the legs slightly, saunter forward with thumbs hooked in belt loops, and then perhaps to mime a tip of the hat to some 'ma'am' in the audience. Someone charged with playing a rapper might move his or her shoulders and hands in a particular way to show this identity. The body is trained how to move according to culture or subculture.

Another place to look for diversity in recommendations for comportment is along gender lines. In most cultures, the genders have different ideals for how to comport their bodies. For instance, some readers' grandparents will have learned that boys bow and girls curtsy when greeting an adult or someone of a higher social status then themselves. Beyond different rules for how boys and girls and then men and women greet people, there are different rules for how men and women should walk, get into a car or onto a horse, and even sit down. For instance, certain cultures hold that women do not sit with legs spread wide—in mainstream U.S. culture, for instance, that mode of sitting is reserved for men.

Embodiment

While other groups might find either mode of sitting strange, the point here is not so much the exact recommendation but that there are recommendations to start with—that there is a 'right' way to sit and a 'wrong' way to sit. Of course, the well enculturated man or woman will sit how he or she will sit without thinking about it as having been taught. Culture is habitual and often out-of-mind this way. Nurtured, cultural modes of comportment thus become thought of and talked about as if 'natural' or given. The damage such biological determinism can lead to notwithstanding, the point here is that modes of sitting (and so on) can become so ingrained or deeply embodied that we do not even realize how cultural our actions are. This, in fact, is one definition of **embodiment**: the literal 'making physical' of culture.

Continuity and Change

Watching someone sitting like a lady, or like a gentleman, also drives home the way in which culture is created in the practice or performance of it. Every time that a 'lady' sits properly she performs as per her culture. If she has paid attention to what 'ladies' around her have been doing—if she has studied her culture—her performance will go very well.

At the same time, if enough people refuse a cultural recommendation—if enough people choose to sit in a different fashion—and this could happen when, say, the recommended way of sitting becomes dysfunctional or it stops fitting well with other components of the culture—cultural evolution can occur. This might happen, for instance, and to keep with our sitting example, when the economy changes and women start to do different kinds of jobs.

Biological Predispositions

It is certainly true that some of how we move is determined by the make-up of our bodies. We have hinges in certain places, and we are bipedal. If we were not, we might not need to sit, let alone to learn how to do it. Still, given what we have—given the bodies that are born with—culture plays a key role in the diversity of comportment modes around the world.

Gendered Bodies

As seen in the sitting example, gender is a prime arena for comportment rules. It is a crucial biocultural arena, and one that is taken most seriously in the majority of social settings.

Gender Is an Organizing Principle

In highland Papua New Guinea, people grow yams. Just as Americans have Thanksgiving every November, people in Highland Papua New Guinea often have yam harvest festivals. I bring this up here because it turns out that yams can be male, or yams can be female. Gender is such an important organizing principle that it can even be applied to vegetables.

Some cultures see gender as so important that it is part of the language; each word may be gendered so that even a lamp or a table may be feminine or masculine. Likewise, in many cultures, someone overhearing a phone call may well be able to correctly tell the gender of the person on the other end of the line because gender determines forms of address and other modes of speech. Some cultures segregate men's worlds and women's worlds, guarding carefully against cross-gender contact. Even in our own society, to this day, there are certain arenas left to the women and certain arenas left to the men.

Indeed, gender is such an important organizing principle, even for us, that often the first thing we do when we hear that someone has a new baby is to ask if it is a boy or a girl. We believe that once we have this information we will know how to properly comport ourselves in relation to that infant. We will know what kind of gift to get, or what to do when that baby cries. We have a whole repertoire of cultural reactions that we will draw on in responding to that child based on a perception of its gender. In fact, parents today often want to know if they will have a boy or girl even before birth so that they can decorate the baby's room in a so-called gender appropriate fashion, using blue if the baby will be a boy, or pink if she's a girl.

These choices are, of course, culture-bound—and time bound as well. One hundred years ago, the color choices of Americans would have been the reverse. Pink was for boys—pink being an attenuated or child-friendly version of red, the manly color of power, heat, and action (accordingly, later it was the color of superman's cape). Baby blue was for girls—it being a version of the color of the Madonna's mantle. Of course things have changed, but in any case, and whatever color scheme might be *au currant*, our comfort level goes up if we know which sex—and so (we believe) which gender—to expect.

Gender Is Not Sex

We have established that gender serves as a key organizing principle for social life. It is what many of us think we will learn about when we ask a baby's sex. We may be very wrong, experientially. We definitely are wrong, technically speaking.

Sex, technically speaking, is a biologically differentiated status. Sex has to with genitals and chromosomes. Sex generally has been determined by looking and by seeing and ascribing meaning to whether a person was born with a penis or a vagina. Today, we also can use genetic tests.

Gender, in contrast, is a cultural construction. It defines masculinity and femininity for us. Gender variously entails expectations of how a person will move, dress, modify the body (for example, by shaving certain parts of it), and so on. It variously entails ideas about what a person should eat or drink (steak or salad, beer or wine, a sloppy Joe or quiche Lorraine), and about appropriate labor or professions. It may entail all these things and more, or only some of them, depending on the culture.

Sometimes gender even entails assumptions regarding to which sex an adult person should be attracted. Mainstream Americans believe that masculine men are attracted to the female sex, and feminine women to males. Historians such as Jonathan Katz (2007) relate the focus on heterosexuality per se to falling eighteenth century middle class birth rates in the West and related demographic worries. Prior to that, reproduction and lust were, in theory at least, quite separate matters. As the heterosexual imperative caught on, and in similar contexts in other cultural settings, homosexuality threatened to undermine gender claims just as heterosexual exploits fortified them. Gender is not always so tightly linked to sexual expression, however; this differs cross-culturally and, it seems, with demographic trends. The looser the link, the less notice people take of homosexuality and so on.

Likewise, gender (the cultural construct) and sex (the biological entity) can exist to a large degree independently. In such a context, if they become uncoupled, it does not cause such a fuss. Accordingly, we must define sex and gender as separate entities if we are to use them to make intellectual progress in unpacking the hows, whats, and whys of gender diversity.

Opposing Examples

While most of us today can understand, based on our experiences, that sex and gender are conceptually different things, this was at one point in time a revolutionary idea in the United States. Not until Margaret Mead published *Sex and Temperament* (2001 [1935]) and *Male and Female* (2001 [1949]) did people really start to come to grips with this idea. Prior to that, more clung to the notion that gender expressions stem from the natural propensities of male or female bodies to act in certain ways.

Margaret Mead was a student of Franz Boas who, as shown in Chapter 4, stood very much against racism and the ethnocentrism it represented. He countered biological determinism with cultural determinism, which puts causality on the side of culture rather than nature. In this, cultural determinism is of course wrong: both nature and nurture are important and in fact they are conjoined. Nonetheless, Boas provided an alternative to the biological determinism rife in the late 1800s and early 1900s.

Margaret Mead applied a culturally deterministic framework in her research, too. One of Mead's major early projects was to counter the biologically deterministic point of view that most people in the West had about gender. What Mead did was ask if the same beliefs that U.S. people had about how males and females should and should not act, and practices thereof, were found across all cultures. If they were found universally, then perhaps there was some basis to the notion that sex begets gender, or that gender naturally arises out of sex. On the other hand, if beliefs and practices differed cross-culturally, then the hypothesis that sex determines gender would have been falsified or disproven.

To test the hypothesis, Mead went to New Guinea to do ethnographic research. The resulting book was widely read by the literate American public. It reported that, in the three cultural groups that Mead visited, there were some very different gender configurations. Mead found that among one group (the Tchambuli), men actually acted in a way that most folks in the United States at that time would consider quite feminine. In other words, Tchambuli men took care of children, were concerned with social relations, and so forth. Women there acted in a managerial and undomestic way that we would consider quite masculine. In a second group, the Arapesh, everybody—male and female—acted in what we would call a feminine manner. In a third group, the Mundugamor, everybody acted in what we would call a masculine way. These differences were highlighted by Mead to demonstrate the culturally constructed nature of gender. Mead's findings would have been even more interesting had she not imposed her U.S.-derived comparative framework; she might have seen, then, that other cultures can have wholly different expectations—not just ones that reverse

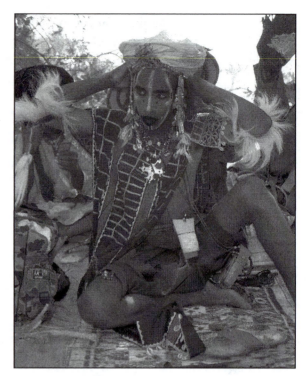

Figure 11.3 A Wodaabe man preparing for a Yaake demonstration; Niger, 1997. *Photo by Dan Lundberg via Wikimedia Commons, CC-BY-SA-2.0.*

our own. Nonetheless, her work did make clear that gender conformity is culturally relative. A woman who wanted to stay home with the children among the Tchambuli would be bending gender expectations; she would be an anomaly in the cultural context of her world.

Another example that turned expectations upside down came from among the Wodaabe, a nomadic cattle-herding group that traditionally has lived in the African Sahel. Among the Wodaabe a time comes when all of the nubile marriage eligible men parade themselves in front of the women, showing off their beauty in a way that might remind us of the Miss Universe competitions. The men preen for hours in front of mirrors prior to their performances; they take great care with their hairdos and apply cosmetics to ensure that the women's highest male beauty expectations are met (see Figure 11.3). The women, for their part, get to choose in the evening which men they will take home with them that night.

Despite this, Wodaabe men are all much more physically fit than many of men in the West today. This is due to the active Wodaabe lifestyle and the lack of labor saving devices in their world, as well as due to the nature of their masculine labor—labor with which they, therefore, shape their bodies as they live. So in many ways, and on average, Wodaabe men meet the West's stereotype of masculinity much better than average Western men in sedentary professions can do.

Biological determinism in regard to gender holds no water. Sure, there are some features of gender in most cultures that do link back to biology—that can be tracked in terms of sex-linked genital, mammary, and hair growth pattern differences, for instance. Yet there are large swaths of gender that do not overlap with sex. Cross-cultural differences in the ways people enact masculinity and femininity demonstrate this. Some readers may not be immediately comfortable with the fact that Woodabe men wear makeup, or that Tchambuli women so firmly occupy the public and not the domestic arena. Nevertheless, all readers should be willing to accept this relativistically as part of the diversity of gender across cultures.

Anomalies

Androgyny Saps Gender's Value as an Organizing Principle

When it comes to subverted gender expectations in mainstream American culture, even staunch relativists may not be so sanguine. One reason for this goes right back to those babies whose sex (really gender) we wanted to know. In mainstream America, without insight into someone's gender role, we find it difficult to organize our social interactions.

That is exactly why the *Saturday Night Live* skit with Pat was so funny. Even the reader who has not seen the Pat skits will understand clearly what the skits were about when I say that Pat was **androgynous**—neither clearly male nor clearly female (the term is derived from Greek: 'andr' means man and 'gyne' means woman; the suffix 'ous' means possessing or full of). The Pat skits were funny because they showed so clearly the social discomfort that people felt when unable to pin a gender onto another person. In refusing to fit one standard or the other, Pat enacted a **transgendered** role: Pat crossed or straddled the border between male and female ('trans' is Latin for across or beyond; while androgyny focuses on a blend, trans roles in some ways leave our two-gender system behind). This flummoxed Pat's coworkers, who were so used to using gender as a cue to social interaction that in Pat's presence they fumbled about, feeling confused and embarrassed. The audience laughed not at Pat, but at them.

In anthropological parlance, Pat was an **anomaly**—Pat did not fit into preconceived, categorically separate gender ideals. Note here that I keep saying 'Pat': our culture's insistence on gender is so deep that I lack a neutral pronoun.

Beyond Pat, another example of the gender-anomalous individual in the U.S. context is the female body builder. Take also the male cross-dresser or transvestite—the man who wears dresses. We may generate all kinds of moralizing rhetoric in trying to explain any anxiety that the transvestite or female body builder's existence might provoke for us. Such effort really is not necessary, however, because the explanation is quite simple: such individuals subvert or challenge the existing theoretical social order. They cross or bring together what we have been taught, culturally, to think of as separate.

Our body ideals exist in part because of the way that they reinforce the culturally preferred social order. Another way to say this is that a culture's preferences for social order are recapitulated in that culture's preferences for bodies. So, for example, if a person comes from a cultural world that prefers to keep male and female or masculinity and femininity separate, a body that overtly brings them together is a body that does not support the illusion (or culturally-agreed-upon convention) of their categorical separateness. In this, such a body makes members of that culture very nervous indeed.

Anomalies Are Powerful

Any anomaly draws its power to discomfort us from the simple fact that it does not fit into the given categories that make up a particular cultural world. For instance, steak-flavored ice cream would be an anomaly for the mainstream American. Likewise, people who used only to be able to find work in circus sideshows were anomalies.

Anomalies both attract and repel us; in this, they have power. Their power has its roots, some theorists say, in the human penchant for order. We approach the world with a preconceived notion of what the orderly life should consist of, with everything in the right place. A thing not where it is supposed to be—"matter out of place," to quote Mary Douglas, a social anthropologist who has thought a lot about these things (1966)—has power because it disturbs, or represents a disturbance in, our very sense of order. In Douglas's view, the key model for order in life comes from the social structure.

Whatever the ultimate source, the proximate or immediate power of the anomaly inheres in its anomalous nature. The mandrake root, featured in Harry Potter's story and often used in magic or sorcery and in traditional medicine, is powerful because of its form (see Figure 11.4). Mandrake roots look human, as if they have legs and arms. In this, unlike the simple carrot, the mandrake seems to cross or bring together the world of the human and the world of the plant. It draws its power from that simple fact—from its anomalous, category-crossing nature. Many plants used in magic or medicine have crossover characteristics, whether in color, smell, or shape.

Anomalies have power. What that power can do, and whether it is considered sacred and pure, or profane and polluting, depends upon the ffruct mandragore.

Fructus mandragore. opto. fri. mi. fic. i i. Electro magni coloxifen. unnani. odoraco ora fod.i.
calam. 7 ugilias. emplauto defenie 7 ifecco oib: nigris euns. nocum. obetat fensus. Re
necti. cū fructu exere. Quid euar no e comestibile 5 ouente. ca. iuu: ibi estate 7 mediamis.

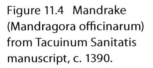

Figure 11.4 Mandrake
(Mandragora officinarum)
from Tacuinum Sanitatis
manuscript, c. 1390.

cultural context within which it is interpreted, and how it is handled. Take, for instance, a chicken with two heads. In some cultures, that chicken would be revered as a saint, or seen as a harbinger of good luck. In others, it might be considered polluting, and people would sooner not see it: it might harm them, for example, by causing their own children to have two heads. In this case, contact with the chicken—let alone eating it—would be **taboo** or strictly prohibited.

While we are considering anomalies that might be eaten, here are a few more examples. In some cultures the dolphin's nature as an air-breathing mammal that lives in the water when mammals 'should' live on land makes it taboo as a food. An egg with an embryo in it rather than just yolk and white, and which therefore brings together death and life, may be taboo as a food; in other contexts it may be considered a delicacy because of this. Importantly, in cultures where such an egg is eaten, certain ritual precautions generally are taken to ensure that its power is productively directed. For instance, the egg could be eaten as part of a fertility rite. If eaten without proper precautions, it might beget octuplets—or lifetime infertility.

Handling Anomalies

An anomaly's power must be handled with care. Oftentimes, those entrusted to deal with anomalies are specially trained priests or shamans. In the *Harry Potter* series, students in the first few years are not proficient in executing magical spells or controlling magical items or substances because they have not yet received enough needed training. Concurrently, at Hogwarts, or anywhere else that magic is practiced, anomalous and powerful items or persons often are carefully stored or otherwise ring-fenced and held separate or kept contained. They and their supernatural powers may be the focus of worship and supplication or subservience.

The concept of anomaly and the related issue of power (whether holy and purifying, or polluting) make it easier to understand why some people find disturbing a body that mixes some characteristics defined as masculine and some defined as feminine. It also explains how and why, in some cultures, such bodies, as well as other anomalous body types (the deaf and blind, those with extra fingers), are respectfully treated. People with anomalous bodies may function as a link from the natural to the supernatural world. Indeed, awe and respect is exactly the response that many cultures have to people like Pat, or to men who would be feminine or women who would be masculine.

Sex and Gender, Unlinked

A Zuni Example

In many cultures, the overlay of one gender on the 'opposite sex' does not raise eyebrows because sex and gender are not tightly linked. Let us take, for instance, the Zuni people, who live in the Southwest United States and have been the subject of a good deal of work on gender because of the existence of the cultural role of the *lhamana*—the male who dresses and works as a woman—the 'two spirit' person (see Roscoe 1992). The lhamana crossover role is not only culturally available but also culturally valued and, in fact, necessary in cultures that have it. The lhamana serves an integral productive function. In fact, in some cultures, if the lhamana role (however that culture terms it) is not filled voluntarily for a village or group, someone will be recruited.

In our Zuni example, the fact that sex and gender are conceptually decoupled is seen in the way that, traditionally, a child is not addressed as a he or she until the child chooses the gender that he or she would prefer to live by and express, for example, in the clothes that he or she chooses to wear or the labor that he or she chooses to do. For instance, a Zuni male child who likes to do the labor of weaving—feminine labor—would be choosing to live as a she (see Figure 11.5).

Conceptual Possibility

The possibility of taking on a two spirit or crossover identity is much greater among cultures that value ambiguity or see the natural world and the supernatural world, for example, as vitally conjoined. In the cosmologies or religious systems in which transformation and ambiguity are central themes—where, for instance, gods or spirits interact with humans, transforming themselves into various living things with various sexes—the crossover role is much more likely to emerge as a culturally sanctioned choice. It also is likely to serve a necessary social role.

Functions of Gender Crossover Roles

Cultural Conservation

Gender crossover roles have a number of functions. These include (1) economic, (2) spiritual, (3) socially mediating, and (4) role modeling functions.

Figure 11.5 We-Wa, a Zuni berdache, weaving. Photo by John Hillers, c. 1871–1907. *Smithsonian Institution, Bureau of American Ethnology, National Archives no. 3028457.*

To begin, if the sex ratio is somehow skewed in a society, and it benefits one sex to take on the labor of the other, cultural notions of masculinity or femininity could change to include those types of labor. Another way culture can evolve is to provide a crossover role in which economic needs might be fulfilled.

For example, in cultures where families need male heirs, it is quite common for a female, if there are no brothers, to take on a male gender identity to support the lineage. Albanians, for example, have a labeled, culturally provided masculine role for females in this position.

This direction of crossover (females acting as men) is much rarer cross-culturally than the other (males acting as women). This may be because so many cultures are patriarchal, where the feminine role is considered a lesser role when compared to the masculine one. In such a context, to see a female take on a masculine role seems quite logical. As manhood is perceived as better to begin with in the eyes of members of a patriarchal culture, this type of crossover is simply more expected and less remarked upon—and therefore less culturally developed—than its opposite. Other theorists submit that the masculine identity is harder for little boys to internalize, and therefore more males are interested in acting feminine than vice versa. Whatever the reason, recognizable, respectfully labeled, culturally sanctioned, culturally available roles for feminine males are more common than those for masculine females cross culturally.

Beyond economics, another and perhaps more important group wide function of gender diversity is to introduce into a given group at least one person to serve as a spiritual functionary. As alluded to earlier, cross-gendered, two spirit, or transgendered people can provide spiritual services due to their anomalous nature.

The spiritual functions of gender diverse people run the gamut. They can work as high priests or shamans to keep the supernatural and natural worlds in balance, ensuring the continued survival of their people. On a more personal level, but ultimately reflecting the same need for continuity, someone having trouble conceiving might call on a two spirit individual to conduct certain rituals to help her to conceive. Other rituals can be performed to insure against crop failure or exam problems or even to help assure success with a love affair or a bet on the horses.

A third function of someone in a crossover role is to serve as a go-between or a mediator for men and women here on earth. A two spirit person could broker heterosexual marriages or carry messages between boyfriends and girlfriends.

Last, but definitely not least, gender diversity provides role models. Commonly, the individual who chooses to take on the labor and dress of the opposite gender does so in a way that puts those who are 'naturally' of that gender to shame. For example, take We'wha, the famous Zuni two spirit person pictured in Figure 11.5. We'wha's weaving simply was the best: it was weaving that any girl or woman could learn from. Likewise, an Albanian female who took on a masculine identity to become the heir in her family could shoot a rifle and hold her liquor better than any man near or far.

We can see this pattern in certain cross-dressing performers today. RuPaul, who gained fame in the 1990s, was quite conscious of the over-the-top nature of the RuPaul persona's femininity. RuPaul said at the time, "When people say 'you dress like a woman', I say 'I don't think I do because women really don't dress like this'. I dress like our cultural made up version of what femininity is, which isn't real" (Harron 1995).

Unreal Ideals

The unreal nature of some of these ideals is seen when we compare the Barbie ideal to reality. For instance, according to the Centers for Disease Control and Prevention (2008), the average American woman weighs 164.7 pounds, and stands just under five feet four inches tall. As per the clothing industry (Zernike 2004), women who might best be compared to Barbie (white women ages 18–25) wear a size 14 dress and measure 38 inches around the chest, 32 inches around the waist, and 41 inches around the hips on average. Non-white women are, on average, somewhat larger. In contrast, if she were real, Barbie would be decidedly taller and lighter, with top-heavy yet still more hourglass-like measurements of 39, 18, and 33 inches (McDowell, Fryar et al. 2008). Incidentally, true hourglass figures, popular especially in the mid-1800s, were achieved through corsets and tight lacing, which could deform the ribcage and organs (recall Figure 0.2, in the Introduction).

Not only females have a hard act to follow. Men also are under pressure to live up to cultural ideals, for instance in regard to their hair (see Figure 11.6). Male toys have hard-to-achieve body shapes, too—and increasingly so; for instance, Jackson Katz notes that GI Joe's bicep measurement equivalent went from 12.2 inches in 1964 to 26.8 inches on the 'extreme' version in 1998 (as interviewed in Jhally 1999). Katz argues that the widespread violence in the United States, including youth-initiated school shootings, is linked to the crisis in masculinity that such inflated measurements reflect and promote.

Figure 11.6 Ad for the Modern Vacuum Cap Hair Grower, as published in the July 18, 1909 issue of the syndicated *Sunday Magazine*.

Figure 11.7 Original movie poster for the film *Some Like it Hot* released in 1959.

Cultural Critique

We internalize many images and ideals whole hog, and cross-gender behavior based on such ideals can and does reinforce them, supporting cultural conservation. Still, some forms of gender diversity openly question or even critique cultural norms.

Take, for example, the motion picture comedy *Some Like it Hot*, which concerned two men (played by Tony Curtis and Jack Lemmon) who got themselves in trouble and so hid out by traveling with an "all girl band" that happened to include a character played by Marilyn Monroe (see Figure 11.7). Compared to the 'real' women in the band, Curtis and Lemmon (or anyhow their characters) simply could not pull womanhood off, and that was the joke of the movie. In this case, the cross-gendered men served as anti-role models, reinforcing the need for men to stay manly (Wilder 1959).

On that basis, you might think I should have included this example in the 'cultural conservation' section above. Some theorists do focus on the pro-patriarchal, misogynistic (anti-woman) sentiment exhibited in this kind of crossover act. Certainly, there is some misogyny in the film. But there also is something else going on. In some ways, the gender diversity of *Some Like it Hot* was subversive; it questioned mainstream norms.

In the film, a man (Osgood, played by Joe Brown) falls in love with a man dressed as a woman ("Daphne") but a man nonetheless (as played by Jack Lemmon). That in itself was no big deal. There are plays from Shakespeare's time and before, and in non-Western cultures, too, in which this kind of thing happens, exposing the constructed nature of gender.

What is different here is that for Daphne the 'misunderstanding' does not get resolved in the way that it does in a culture-conserving comedy of errors. Instead, at the end of the film, Daphne and Osgood get into a motor boat, with Osgood intending to speed his little lady away into the night, and eventually to wedded bliss. Osgood says something like, "I called Mamma. She was so happy she cried. She wants you to have her wedding dress." Then Daphne replies, in falsetto, "Oh, Osgood! I can't get married in your mother's dress." Osgood says, "Oh, that's all right, we can have it altered." Daphne says, "Osgood I have to level with you. We can't get married." So Osgood says, "Why not?" Daphne says, "Well I'm not a natural blonde." "Doesn't matter," says Osgood. "I smoke, I smoke all the time." "I don't care." "I can never have children." Osgood here says, "We can adopt some." Finally, completely exasperated, Daphne says, "But you don't understand." Daphne rips off her wig, drops the falsetto, and says, "I'm a man." Osgood remains smiling, driving the motor boat like nothing happened: "Nobody's perfect," he says with love's contentment (Wilder 1959).

The film thus not only calls attention to the constructed nature of gender, but threatens to undo it. This was why that film was considered so subversive at its time that its posters were stamped "not suitable for children." In sum, gender diversity is not always conservative in its function; it does not always act to preserve or reinforce existing expectations. It can also be very subversive, depending upon how it is deployed. While in some contexts flouting norms may be punished, in others it can lead to culture change.

Rites of Passage

Constructive Function

Until this point, we have focused on differing cultural expectations, not only for what is masculine and what is feminine, but also for how tightly or how loosely sex and gender are coupled to begin with. We've explored how people shape, train, and comport their bodies as per gender expectations, which differ cross-culturally. The argument that culture constructs gender, and in doing so impacts the body, gains further reinforcement when rites of passage are considered. In many cultures, throughout time and around the world, it takes a very specific rite of passage to turn a person from a boy into a man or from a girl into a woman.

What Is a Rite of Passage?

A **rite of passage** is simply a set of ritual acts intended to move a person or a group of people from one social status to the next. The first haircut is in many cultures a very important rite of passage to bring a baby either from being a non-human organism to being human, or from being a gender-neutral baby to being a boy or a girl.

Some Jewish groups, for instance, see the first haircut as an important rite of passage for little boys. With the first haircut, which comes at about the age of three, the boy is deemed ready for the next phase in his journey to manhood. I should note here that this rite of passage is actually one in a series of rites of passage that altogether create a person into a man. The little boy has already undergone a circumcision, as an infant. Now, as he grows, he will every few years undergo a new rite of passage moving him, through a series of social statuses, further and further toward manhood. In each rite of passage, other community members participate in helping to transition the lad from one status to the next. For ultra-Orthodox Jews, a boy only can be created into a man through the concerted efforts of men.

Social Birth

In ultra-Orthodox Judaism, delivering a male baby from one's womb is just a precursor to the boy's 'real' birth, which happens when he gets circumcised one week later. Technically, his father must do that, and, technically, it is not just a clip of the foreskin that is entailed. Circumcision is part of a long process of uncovering the real man within. When a fetus is in the womb, it is in the amniotic sac and sometimes, when babies are born, bits of that sac actually can adhere to them and must be pulled off. The circumcision is a continuation of this kind of unsheathing. So, in effect, the father who circumcises the boy is the one who is the one who delivers him, socially speaking.

Recently, as described by Yoram Bilu (2003), ultra-Orthodox Jews have brought together the first haircut and school initiation. This cultural adaptation happened in the years after World War II, partly because of increased pressure to assimilate. One idea behind doing school initiation earlier and combining it with the first haircut was to get the little boys into school sooner. Because they do go to their own schools, this was a way of protecting the boys from the influences of the broader world. Of course, the cultures of other Jewish populations have evolved in other directions. The ultra-Orthodox example is just one of many.

In this community, these two rites of passage have come together so that each little boy is simultaneously separated from infancy and brought into the male world of learning, and removed from the female domestic domain. Mother's breast milk, rabbis say, is now replaced by the sweet nourishment provided by learning. Accordingly, during the rite of passage, the rabbi who replaces the mother holds the young one on his lap. He puts honey onto a special piece of paper that has all the letters of the alphabet written on it, or writes the letters out himself using honey for 'ink.' He then offers the page of sweet letters to the boy who licks them up in lieu of breast milk. In the ultra-Orthodox series of rites, creating a boy into a man involves not only uncovering him first through circumcision and then a haircut but also through introducing him to letters and words—to the world of knowledge, or of the book—and separating him from his mother.

This rite of passage is orchestrated and witnessed by men and boys in the new initiate's community. It is a sweet occasion, and the boy's release from his coverlet of hair as well as the opening of his eyes to learning are commemorated by sharing (what else?) a honey cake. All the boys and men get a bit of the cake and thereby not only share the sweet experience but also incorporate it into their bodies through the process of ingestion. The act of shared eating, called **commensality**, is done in many ritual contexts; it is done whenever people ceremonially break bread together. Two people who have eaten from the same loaf cement, through that shared substance, their social bond to one another. This kind of sharing will happen again for this little boy at his Bar Mitzvah and then at his wedding and of course when he circumcises his own sons; family building is a key goal of manhood in the ultra-Orthodox Jewish tradition.

Boy to Man

Across cultures, there are more rites of passage bringing boys into manhood than girls into womanhood. Some theorists explain this as due to the female body's capacity to menstruate, which provides a natural marker of the onset of physical womanhood. Other theorists, particularly those with psychoanalytical training, believe that the journey to manhood actually can be harder because, in most cultures, women raise children.

In cultures that have constructed highly distinct and segregated male and female spheres, in which the men occupy a completely separate world from the women and vice versa, it is much harder for little boys to become

Figure 11.8 Nine- to ten-year-old boys of the Yao tribe in Malawi participating in circumcision and initiation rites. *Photo by Steve Evans from Citizen of the World, CC-BY-2.0.*

men because it is a longer jump for them to make. They are not simply moving from one sub-status to another sub-status in a gender-integrated world—one in which they have male role models to interact with and study all the time. In gender-segregated cultures little boys, as children, stay with female kin. Because men often stay away, living separately, the boys have little idea what men do. While girls see women all the time, boys growing up in a gender-segregated context may find masculine identity formation truly difficult, at least initially.

Elaborate rites of passage for boys in their young adolescence help sort this out (see Figure 11.8). Indeed, male rites of passage in these societies often are intense and long. The series of trials they involve can include feats of physical strength and endurance or resolve. While Jewish youth undergo circumcision as babies, in other cultures this is done when boys are teens. Sometimes, more intense genital surgeries are required (the hazards of which may be seen in the epidemiological profile of the group implicated). Certain groups in Melanesia, Australia, and elsewhere, perhaps particularly those that are very gender segregated, practice subincision, in which the penis is split on the underside from the urinary opening. Other groups require tattooing or the like during a rite of passage. Still others require a boy to kill a large mammal or to experience a vision or to demonstrate his worth by other means.

Because the body is central to this book, let us explore in a bit more depth some of the cultural understandings about the body that are worked through in male rites of passage (also see Burton 2001). In some cultures, boys raised by women are understood as thereby filled with female qualities, in part due to drinking mother's milk. To create them as men, that milk must be replaced, as it was with the sweetness of learning in the ultra-Orthodox Jewish school initiation rite. Other people replace it with other things, including sometimes male semen, a masculine life force.

This is perhaps most famously done among the Sambia, whose initiation rites Gilbert Herdt studied (1994 [1981]). In this culture, and some others, boys perform oral sex and thus feed on their male guardians as part of their rite of passage. Ingestion of semen and its biological incorporation into their bodies helps make them men.

Among the Sambia, as in many other societies, the boy-to-man transition is a group affair. This highlights the boys' importance to the community and helps reinforce community bonds—not only among the elders conducting the ceremonies but also among the initiates going through them. Children who undergo their rites together have a very strong bond with one another for life.

Group rites also have practical importance in a more mundane way. These ceremonies are not only intense in terms of content—they also are time and resource intensive. It is more economical—in terms of time, effort, and resources—to ferry a group of children through a rite of passage together.

Structure and Process

Every rite of passage has three main parts, as represented in Figure 11.9. It always begins with an act of separation in which an individual or group of individuals are separated from their initial social role. This is represented at the junction of the first horizontal and vertical lines, with the number "1." The lower horizontal axis represents the mundane or everyday world, and the higher one represents sacred time and space, during and in which the second, transitional (liminal) phase happens. Rebirth (reincorporation) is phase three, and the end of the rite of passage.

Figure 11.9 Three part structure of Rites of Passage: 1 = Separation or pre-liminal phase; 2 = Transition or liminal phase; 3 = Reincorporation or post-liminal phase *(see Leach 1979 [1961]; Van Gennep 1960 [1908]; Turner 1979 [1964]).*

2. Liminal Phase

SACRED TIME

1. Separation

3. Reintegration

MUNDANE TIME   MUNDANE TIME

Separation

To begin a rite of passage, the initiates have to be separated from their old social position. Essentially, a switch must be thrown. While this may happen as gently as being awakened with a sweet song not sung on any other day, sometimes this happens in a most frightening way for the children or child involved. It might happen at night, when slippage between the everyday world and the supernatural realm is most likely. A child could be sleeping peacefully with his mother one moment and be kidnapped, so to speak, the next by male elders or the spiritual leaders of the group—people with whom he often has not spent much time and who may be disguised anyhow (I describe rites of passage from the male perspective in keeping with the points made in the previous section; but of course females go through them, too).

Ripped from mother's bosom, the child may be stripped of any regular clothing, perhaps even of his hair, and taken to a spot far from the village. Only when an individual is stripped of or separated from a previous status can he assume the next. Death symbolism is prevalent during the separation process.

Transition

The second, transitional stage is famously termed the **liminal** period. This label we have inherited from Arnold Van-Gennep who, along with Victor Turner, greatly enhanced our understanding of rites of passage (Van Gennep 1960 [1908]; Turner 1979 [1964]). The label comes originally from the Latin word for threshold, or for the small area dividing one room from the next, or the inside of a house from the outside. A person going through a rite of passage is, in effect, standing in the doorway between social statuses.

Once our boy's old status is gone and he has been, in effect, removed from the secular, mundane world and its every-day social structure and trappings, he comes to occupy 'inter-structural' time and space. In Figure 11.9, this liminal arena is represented by the higher register on which this phase is depicted. The person in transition here is "betwixt and between" social statuses (as Turner famously put it). As such, he occupies a powerfully anomalous, sacred space.

Recall that the power of those who cross categories or do not fit into them must be handled with care. Oftentimes, the initiates are therefore sequestered: the mere sight of them might harm regular village members—or vice versa. If this sounds farfetched, consider that in mainstream American marriages it has been considered bad luck for the groom to see

Figure 11.10 Platoon Sergeant Shawn D. Angell trains marine leader candidates at the Officer Candidate School, Marine Corps Base, Quantico, Virginia. *Photo by John Kennicutt, U.S. Marine Corps.*

the bride on the day of the wedding; often he does not see her face until after vows are uttered and her veil is lifted.

Just as the bride, when getting dressed for her wedding, might be given special instructions for how to be a success as a wife by her mother or other female relatives who already have undergone that particular rite of passage, so, during their confinement, are boy initiates taught what they need to know to be men. While being treated as liminal already heightens awareness or arousal, many cultures induce altered states in initiates to help their minds grab onto key information or to open them up to the change that must be undertaken. They may be kept up all night for a number of days; they may be forced to fast; they may be asked to sing for hours on end, or to participate in drumming. Pain and fear also can be used to reinforce the transformative effect of lessons learned in this phase of the passage (see Figure 11.10).

Something else undertaken both to speed up and enhance learning is symbolic communication. Symbols packed full of cultural meanings related to the transformation that is being orchestrated are deployed. Often, such symbols are monstrous combinations and they can be very scary. The combinations may emphasize qualities that come together in the next status that the individual will be passing to. For instance a statue of a man's body with a lion's head may be used to emphasize the union of brute physical strength and the power of the intellect in the grown man. This head and body combination represents an inversion, because lions are stronger than men and men smarter than lions—but the message lodges in this very topsy-turvy fact. Symbolism in rites of passage is generally topsy turvy, being presented within the crazy, mixed-up, upside down world of liminal, intra-structural time and space.

Also and accordingly, in these ceremonies, death and life images and male and female ones often are depicted in combination. Male or female images will have exaggerated features, such as impossibly huge genitals or dozens of breasts. While being shown the various symbols, often via ritual objects such as statuettes, or in costume decorations, the initiates generally hear stories explaining, explicitly or implicitly, what is and is not proper behavior for someone of the status that they are about to earn. The initiates' heightened arousal gives messages of extended intensity and staying power.

Largely due to the intensity of rites of passage, people going through them together form a bond of community with one another, called **communitas**. One might be the son of the king and another the son of a sharecropper, but in a rite of passage, every individual is of equal status—because nobody has any status whatsoever. Initiates have been stripped of that (see Figure 11.11)

Figure 11.11   Basic trainees rally themselves with motivational chants while waiting to navigate the obstacle course at Lackland Air Force Base, Texas. *U.S. Air Force Public Affairs.*

Figure 11.12 Corset advertisement from a 1934 catalog by National (a mail-order company). *Courtesy Katharine Zenke (anadaday. wordpress.com).*

The loyalty that going through initiation trials together creates can persist to old age. Some cultures formalize these ties, giving each group of initiates or 'age mates' a name and by assigning these groups certain ritual responsibilities. Socially, these groups can help bind a village or community, providing ties that cross-cut and so knit together disparate lineages.

None of what I have described is totally unique to non-Western cultures. In the United States, we have fraternity and sorority rushes, complete with sometimes quite dangerous hazing rites. Boot camp entails the elements of a rite of passage. Through humiliation, degradation, pain, rituals that call for endurance, and the sheer experiential intensity of basic training, civilians can be transformed into military women and men.

Reincorporation

At the end of the transitional or liminal phase, when the elders have deemed the initiates sufficiently educated and sufficiently transformed to take on the rights and obligations of the status that awaits them, they are reincorporated into society. The symbolic imagery used here is frankly that of rebirth. In other words, some of the same rituals undertaken when a new baby is born also are performed when initiates emerge to occupy their new status. In one instance, in a New Guinea group, freshly initiated young men being reincorporated actually crawl back out into their social world through a tunnel or birth canal created under the spread legs of their male elders. This is very much like the social birth that happens in Judaism with father-driven circumcision.

So men make men. Likewise, women make women in female rites of passage that do exist. Long or short, individual or group-oriented, intense or fairly easy, men's and women's rites of passage work to create properly gendered beings. What is proper varies across cultures.

Gender: A Key Site of Biocultural Interaction

Gender, far from being simply a natural fact, is a key area in which biology and culture come together. It is created in diverse ways in different populations through the embodied practice of culture. It is put into play using bodies that are sculpted in particular ways through population-specific cultural practices—sometimes quite knowingly so (see Figure 11.12). We will continue to examine the junction of biology and culture in the next chapter, which looks more closely at how culture, itself a biological imperative for humankind, in turn affects our biological form and function.

Dynamics of Human Biocultural Diversity: A Unified Approach by Elisa J. Sobo, 233–262.
© 2013 Left Coast Press, Inc. All rights reserved.

12

Variation in Body Ideals and Outcomes

This chapter prepares you to:

- Review examples of body decoration and modification cross-culturally, with an eye to their biological impact

- Explore cross-cultural and historic differences in weight ideals in relation to food supply

- Examine weight-related health challenges in light of our evolved metabolic biology

- Identify and describe the functions of body modification and decoration cross-culturally, and discuss the main dimensions along which such practices vary

Ideals for the body, including its height, weight, and shape, and the ways in which it can be decorated, modified, and sculpted, have been around for a long, long time. Indeed, evidence of our interest in body symbolism dates back to culture's emergence 75,000 or so years ago. Among the artifacts found bearing witness to this transformative event in humanity's history are shell beads and pigments used for bodily adornment. Later, but not too much later, we begin to find tattoo kits, and evidence on human remains of tattooing.

We put culture into practice via our bodies. In part because we do this, our bodies can be treated as texts—as things to be read. When we see a person's body, we make assumptions about that person based not only on his or her apparent sex, gendered mode of dress, or habitual comportment (see Chapter 11) but also his or her height and weight, insignias displayed, the way the hair is kept, and so on. We examined previously how social structure is written into or embedded in the body so that various populations' bodies differ categorically (see Chapters 9–10). Flawless skin and complete sets of teeth, on the one hand, or scars, missing fingers, and even internally inscribed indicators such as hypertension and diabetes or particular forms of bone and joint wear, on the other, can be read for insight into a group's social standing, or a whole society's priorities.

With knowledge of our culture's visual vocabulary, we may even purposefully inscribe ourselves or our loved ones. Body modifications such as circumcision undertaken in rites of passage are certainly one form of such self-inscription, but there are others. For instance, we may become bodybuilders, striving to alter our musculature, whether for aesthetic reasons, to amass strength, or to impress others with our dedication. We may undergo liposuction to conform more to our culture's weight ideals. We may paint our faces and color our hair or wear wigs as football fans to show the world which team we back. As fans, we will want people to be able to read us—to know we support a specific team. Similarly, we may tattoo ourselves or simply label ourselves with a college sweatshirt so that anybody who walks past can read, on our very physical form, something about our group membership.

The embodied signaling we do also conveys information regarding whether or not we endorse the way our society is organized or the cultural values promoted there. For example, wearing a collegiate sweatshirt can convey that one values education, or that one is willing to play by the rules that one's society has regarding how to get ahead, for instance.

Invited to or not, people will gaze upon us. People will read whatever we have to offer them visually. Meaning will be imputed in accord with body ideals that have been established. These ideals vary from culture to culture, and within a given culture, they also vary from era to era. The biocultural context in which particular ideals are created and maintained as well as the knock-on biocultural effects they may have contribute to the range of human biocultural diversity.

Opening Examples

The Model Citizen

Take for example John Bull, an invented figure, who became a mascot for England in the 1800s, somewhat like the United States' Uncle Sam. He often was depicted out-of-doors, in the countryside, with a dog at his feet (see Figure 12.1). He was strong, but also fat, somewhat older, and decidedly middle class.

Figure 12.1 "Conversation across the Water"; John Bull and Napoleon, artist unknown, Vaisseau de Ligne, *Time Life*, 1979.

With these features, John Bull does not fit today's ideals. Nevertheless, in his time, his figure crystallized representationally all that was wonderful about the common English man, with his sensible outlook, emphasis on fairness and generosity, sporting nature, and willingness to take a principled stand. John Bull's inventor, John Arbuthnot, leveraged Bull's body, including its size and how it was dressed, to connote valued traits so that Bull's image stood in for the good social and cultural citizen of England. His body and, by extension, England's body, was fine indeed.

Today, of course, in England as in much of the West, stout bodies symbolize gluttony and greed as well as a lack of self-control or a willingness to sacrifice for the nation. Our model citizen today therefore has a different physique.

The Anti-example

While the body can be used, as John Bull's was, to express pro-social and pro-cultural values, it also can be used to mark and so warn against non-conformity. Non-conformers often have, throughout history and around the world, been **stigmatized** or physically marked as targets for rejection. Their heads have been shaved, they have been tattooed, had their hands cut off, or been forced to wear clothing clearly labeled to let others know that they are not upstanding or otherwise valued members of society. The scarlet letter, for instance, in the book of the same title by Nathaniel Hawthorne, was used to mark its bearer as adulterous at a time when such behavior, for a woman, was completely unacceptable. In the book, Hester Prynne had broken a very important rule, so her clothing—her social skin—was marked to show it and shame her, and to urge others thinking about doing the same to get back in line.

Some stigmata are inherited. For a time in European history, a sunken-looking or small flat nose (for example, with a depressed bridge) was taken as a sign of personal and ancestral degeneracy, as nose deterioration problems were clinically associated with the sexually transmitted disease, syphilis. Cosmetic nose surgeries focused on enlargement. When one saw a person with a small nose, one assumed that person was lecherous and corrupt; one wondered if his or her parents also were (a gap for a nose can result from congenital syphilis). Seeing someone with a small nose also served as a reminder to comport oneself in a way that would minimize chances of contracting the disease (Gilman 1999).

Implications

These examples give us some sense of the way that culture and biology come together in body ideals, with biology providing the raw material and culture the model for how to shape it. The shaping can, of course, have an impact, not only in terms of its superficial results but also its health outcomes, as when eating too much red meat (which John Bull was wont to do) causes heart disease, or a commonly undertaken nose operation leads to nasal impairment, or a stigmatizing mark among members of one group is used by others to support that group's maltreatment. To better understand the biocultural significance of cross-population variation in body ideals, we will take several angles on the issue. After first focusing on overall body size and shape, we will then move on to consider how populations diversely adorn the body, whether through surface decoration or physical modification.

Body Size and Shape

Chapter 11 mentioned how gendered body ideals actually have diverged, in the United States, from what is attainable for the average woman or man. We saw, for instance, how the representations of men and women in toys are not representations found in reality. They reflect cultural ideas about, for instance, men's biceps and women's breasts. Interestingly, these

Figure 12.2 Mauresse, Dowiche by P. D. Boilat, 1853 (from Esquisses Sénégalaises; physionomie du pays, peuplades, commerce, religions, passé et avenir, récits et légendes). *New York Public Library Digital Library 1222721.*

ideals change over time, and they also differ from culture to culture—and sometimes from gender to gender.

Among Azawagh Arabs, a semi-nomadic people who live in windswept western Niger (and who also have been known as Moors), to be beautiful a woman must be fat. Girls are urged and sometimes even forced to stuff themselves with fattening porridge and milk so that they can achieve great physical heft, thereby becoming women. This quickens the onset of menstruation and supports fertility. A swollen body is a demonstration of a proper attitude (see Figure 12.2)

The body of a woman should be 'closed'; just as her stretched skin now contains fat, it too contains the blood and breast milk of kinship. It is crucial that women not diffuse this by sharing it outside of the community; female substance—and their allegiances—must be safeguarded. Big bottoms, stretch marks, and rolls of stomach fat therefore serve both to enhance female attractiveness and to show that a woman is doing her job of containment. It also helps women to do this job by grounding them, quite literally: the larger a person is, the harder it is to move about, and when a woman's role is as an anchor for the family this works out very well indeed, at least symbolically (Popenoe 2004).

There are other cultures that value fat bodies, too, especially for women. In many of the West African societies to which African Americans trace much heritage, those who can afford to do so seclude their adolescent girls in special 'fattening rooms.' After a period of ritual education and heavy eating, the girls emerge fat, attractive, and nubile (see, for example, Simmons 1998). Richer families, of course, can afford to feed their girls well, and to do longer without their contribution to the household labor pool; thus, one's higher status is carried and displayed in one's belly, backside, and other big body parts.

The people of Fiji have traditionally liked fat bodies for both genders. Fatness signifies a wealth of social connections and financial resources and, thus, 'health' (Becker 1995). In fact, the social connections embodied in the fat individual are literal, not symbolic; big bodies are co-created as people feed each other. The popular U.S. idea that weight is a personal achievement is wrong for two reasons: first, social structural factors such as poverty and lack of neighborhood parks foster much overweight and obesity here; second (but no less importantly), weight is co-created by people in social relations.

Weight thus demonstrates the intertwined nature of our bodies. In this, it is not alone. We already have talked, in the context of Chapter 6's discussion of breastfeeding and co-sleeping, about various ways in which infants' and caretakers' (especially mothers') bodies are interdependent. Still other kinds of interdependencies and interconnections exist, as Fijian ideas about fatness suggest. Empathy is one (see Chapter 4). There also are interpersonal biochemical signaling systems, such as those involving others' scents, although a lack of time and space bars me from going into those here. I will, however, take time to note that, in some cultures, a conjugal or spousal couple becomes as one, physically speaking. For example, in some cultures men participate in pregnancy, experiencing morning sickness and other symptoms. The anthropological term for this is **couvade**. Just as fatness implicates more than one person, pregnancy, too, is not a private bodily state but a group one; in this case that group is the parents, together (and see Chapter 13).

Getting back to build, in Jamaican tradition, where a respected adult is called a "big man" or "big woman," good relations involve food sharing, and people on good terms with others are ideally large. I did my initial fieldwork in rural Jamaica; when friends or participants in my research wanted to compliment me, they told me I'd put on weight (Sobo 1993).

Being large is not in and of itself a good thing, however. Jamaicans look and strive for 'good' fat—fat that is firm, not slack. The goal is a body that is plump with vitality. Slack fat signifies decline, as when ripe fruit goes soft. The Jamaican preference is for fat that stands up, as do breasts plumped with breast milk. In this, Jamaican tradition, like most cultural traditions, favors bodies that measure in the United States as overweight—but not as obese. Although there are exceptions, such as with the Azawagh Arabs, extreme fatness (obesity) generally is considered worrisome all around the world.

The opposite of being too fat is being too slim. In Jamaican tradition, slenderness or weight loss signals social neglect. A Jamaican seeing someone grow thin might wonder about the sorts of life stresses that have caused the weight loss, rather than offering congratulations for it and attributing it to a 'good' diet, as many middle- and upper-class Americans do without reflection. U.S. residents living in our inner cities, on the other hand, also worry about weight loss among friends or relatives. Thinness might be a symptom of HIV infection or tuberculosis, it may indicate that a person is addicted to crack cocaine or other dangerous substances, or it may simply signify poverty.

In her cross-cultural review on the topic, Claire Cassidy (1991) noted that bigness tends to ensure reproductive success and survival in times of scarcity and that, cross-culturally, plumpness is considered attractive generally. She also found that socially dominant individuals with sound relationships are usually large (relatively speaking). While in some cultures fatness is increasingly devalued due to the rise in diabetes and changes in diet and activity level that have meant individuals once merely plump are now obese, the long-standing equation between size and health has been harder to shake in areas where famine is still a likely hardship.

A Mismatch between Human Biology and Modern Life?

The recent Western increase in the pervasiveness of food (or things to eat and drink, anyway) represents a major environmental shift for human beings. For most of time, and in most places, food has been a sought after good due not to its deliciousness (although there always was some of that) but for sheer survival. Because food could be scarce, being able to store food energy as fat, and certain nutrients in fat, was selectively advantageous. In other words, our bodies genetically adapted to store at least some sustenance for lean times.

Changes in the food environments of Westernized nations in general, and the United States most specifically, have not only led our bodies to leverage that capacity, packing on the pounds, but also to increase their metabolic capacity to do so. We seem in fact better able now than ever to take advantage of the food that is around us in terms of stored fat.

Calories In, Calories Out

What explains the upward trend in weights worldwide, at least in rich nations but especially—and notably—among the poor there? For one thing, even the poor in rich nations can afford masses of calories on a daily basis. Indeed, the cheaper the food, the more calories it often contains. The poor can therefore afford to eat most of the time, in rich nations, if they are not concerned about nutritional content; they also can get more, calorie wise, for their dollar with such foodstuffs. Who does this benefit beyond the fast food franchises and food producing industrialists? The poor (and others) are fed plenty, but they are not well-fed. The quality of their diet is vastly different, and deficient, when compared to the quality found in, say, the typical forager diet (see Chapter 6) or even an agricultural but pre-industrial one.

Of course, it is not just calories taken in that make us fat; it is also the lack of calories out. Activity levels have dropped dramatically in the fatter nations. Whereas once people here undertook hard physical labor on a daily basis not only to get food but also to prepare it and to keep their homes in order and clean, now labor-saving devices abound. Gone, for most readers, are the days when we churned butter by hand, beat the rugs to clean them, or walked to the river and back carrying heavy jugs of water, burning calories all the day. We just use less energy now in the activities of daily living. Even paid labor often is not physically laborious; many employees sit all day at desk jobs. Too much sitting is, as it turns out, hazardous to our health.

That this is so, as well as that genes play a lesser role in our new roundness than we might like to think, is clearly demonstrated when we compare the body mass indexes or BMIs (a measure of body fat based on height and weight) of the northern and southern Pima peoples. The Pima are a Native American cultural group that includes the Tohono O'Odham described in Chapter 9. Historically, the northern and southern branches were part of the same population, but with the birth of the United States as a political entity those who lived in Arizona gained rights and obligations relative to the U.S. government, while southern members of their population came under the rule of Mexico.

The present diabetes and obesity problems of the Tohono O'odham stem from the unfair diversion of their river and the concomitant loss of their traditional way of life. Without their traditional farming and related foodways and physical activity patterns in play, the Tohono O'odham went from a fit population with no incidence of diabetes to one of the most diabetes prone groups in the United States (California Newsreel 2008a). Their genome did not change; their lifestyle did. They began eating the lard, sugar, and white flour provided by the federal government as a replacement for farmed food. They began using less energy because they were not farming. The infamous Southwestern staple, fry bread, which is just that, is a legacy of this arrangement—not 'native tradition.' Concurrently, the Tohono O'odham experienced the chronic stress of displacement and marginalization (regarding the physical effects of which, see Chapter 10).

In contrast, the southern people, who live in the Sierra Madre Mountains in Mexico, have much lower rates of obesity, overweight, and diabetes. In one study, three of thirty-five had diabetes, in contrast to half (one in two) of their U.S. cousins. A key reason for this is that, south of the U.S. border, traditional subsistence practices remain in place. The southern group's lifestyle is physically rigorous, and their diet retains healthful ingredients (National Institute of Diabetes and Digestive and Kidney Diseases 2002, 5; see also Thomas 1998).

Nobody doubts that the Tohono O'odham who do not lead active lifestyles, and so forth, are vulnerable to what some have termed 'diabesity'; as human beings, of course they are. However, whether they are more genetically vulnerable than members of other populations remains a questionable point. Many populations have been subject to famine but have not gotten nearly as fat or sick. For example, in Ireland, there was the Potato Famine starting in 1845; Russia experienced famine in 1921; many nations embroiled in the World Wars ran short of food. It seems likely, then, that while a genetic component to diabestity must of course exist, the present glut of fatty, sugary, calorie-laden foods in certain environments is the lever that sets the bodily fat storage machine in motion, given certain other conditions such as poverty, unemployment, and chronic stress such as in relation to minority social status.

Metabolic Issues

Food-rich environments—particularly those with industrially produced food—are mismatched to our evolved digestive and metabolic physiology and biochemistry. Many of the ingredients now contained in food are created through synthetic industrial processes, and our bodies often do not have metabolic pathways in place to deal with them. Some use or alter the course of pathways meant for real food components in ways that do not always turn out well.

When we expect to or are ready to eat, we start to salivate, and our metabolism or the rate at which we break down and use or assimilate nutriments picks up, at least under normal conditions. By normal here I mean when the food to be ingested is actual food and not an industrially created substitute. With those, anticipatory responses may differ.

Sugar substitutes, for instance, or foods containing them, may be anticipated as low in calories, which they indeed are, but the foods that often accompany them are not (consider the large fries and hamburger that your friend orders with a diet soda). Anticipating a low-calorie food meal based on its recognition of the sugar substitute in the soda, the body may not rev metabolic processes up to the standard food-digestion rate; a lower metabolic pace in times of shortage helps the body to stretch or make the most of a minimal intake. This is why people on low-calorie diets do, after a time, have lower metabolic rates. The body lowers its metabolic set points to try to regain homeostasis (this is the same kind of allostatic process described in Chapter 10, whereby a change meant to better regulate the body in the short run turns out to dysregulate it in the long run by throwing off other processes).

The problem with having a lowered metabolism when food actually is plentiful, and liberally eaten, is that with a lower metabolism more of the calories taken in are stored than usual (in contrast, the higher the metabolic rate, the more calories are burned). At least this is how things work in rats, whose metabolic processes in relation to sugar substitutes have been extensively studied. The rats also eat more overall when fed a diet laced with sugar substitutes. Scientists do not know exactly why overeating happens, but they suggest that the rats' bodies, anticipating calories from sugar and then not receiving them, send increased appetite signals, pushing the rats to seek out and take in those calories that seemed 'missing.' If these patterns hold true in humans, it may well be that drinking diet soda, for example, makes people fatter than drinking the real thing. Indeed, one study exploring the relationship between diet soda drinking and obesity found that for every can of diet soda consumed per day, the risk of being overweight increased 41 percent (see Park 2008).

It also may be that changes in prenatal exposures have triggered a rise in obesity in some populations via epigenetic mechanisms (see Chapter 3). There is some indication that exposure to certain toxins exuded by particular kinds of plastics predispose a fat-deposit regulating gene to remain activated or to over-express itself, at least in mice, who thereby can become very fat indeed (the toxin is BPA or bisphenol A; the gene, agouti; see Nova ScienceNOW 2007).

Babies born to human mothers whose calories were restricted during pregnancy, whether due to dieting or famine, seem more likely to store fat more efficiently (that is, put on weight easily) as they grow, and they may be more prone to insulin resistance (this leaves more sugar in the blood, which is good if an individual is not receiving much sugar, but dysfunctional in our present, sugar-rich dietary environment, and may end in diabetes). In effect, the calorie-deprived environment that the fetus experiences in the womb of a mother whose diet is restricted is assumed by the fetus to forecast or be a predictor for the environment in which it will live as it grows up: if calories are restricted during gestation, the fetus is primed for a life of scarcity. This can have a negative impact if the actual environment is one in which food is abundant (Bateson, Barker et al. 2004). Cultural beliefs about how much to eat when pregnant, and structural conditions limiting access to food or the reverse, seem to play a role in underwriting a proportion of existing diversity in offspring body size from population to population.

Even when prenatal and postnatal environments match, and even when one eats food to which the digestive system is well-fitted, some aspects of appetite still are humanly created. I refer here not only to the foods on hand due to whichever cultural adaptations have been implemented and wherever in the world one may live, but to flavor preferences determined even before birth (I mentioned in Chapter 6 how breast milk as well as foods eaten during pregnancy can lead to certain flavor preferences in infants and children). So what we eat is bioculturally determined, and variable. When we eat, too, is a biocultural issue: we get hungry around the time that we normally eat due in part to habituation. We also can habituate to the hormone leptin which signals, in effect, fullness, instructing us to stop eating; the vast majority of obese people's bodies give this off, but because they have eaten through or over its signal so often they no longer 'hear' it (Kluger 2007).

Fat Phobia

Even in the United States, plumpness was long considered healthful and alluring. In the early twentieth century, the diet industry focused on helping people gain weight, not lose it. Those who were slender were caricatured as weak and unattractive. When and why did things change? Some reasons are reviewed for us by journalists Kate Dailey and Abby Ellin (2009).

According to their investigations, some scholars trace the rise in fat hatred to World War I. At that time, some Americans believed that a fat population (although ours was far from it at that time, compared to now) would not be able to compete. Moreover, food was in short supply due to the war; fat people were thought to either be hoarding, or to be taking more than their fair share of it. While in some societies where food is scarce fatness is simply taken as a sign of wealth and power, and thereby valued, this could not happen in the context of the war because it was contradicted by the cultural message that we all should pull together and that everyone would have to make sacrifices.

In short, then, emergent ideas about fat were related to ideas about self-control in the face of personal greed or weaknesses of the will. These ideas stuck around when the United States bounced back as a land of bounty. Other developments reinforced this view, such as the rise of sedentary desk jobs as well as labor-saving devices, which meant that activity levels would be low unless exercise was self-imposed. The advent of ready-to-wear clothing, with its numerical sizes, provided people with a ready way to make size comparisons, too. The easier it got to be fat, the more we began to favor thinness. Plumpness, which most cultures around the world and through time had sought after due to what it meant in the face of potential food scarcity, was turned by American prosperity into a fool's prize.

As fat stigmatization has risen in America, people whose bodies are plumper than the cultural ideal are caught in a health bind. On one hand, their risks for certain disease states may be heightened by their overweight status; on the other, if they have internalized the ideals of their culture (as good cultural citizens are wont to do), they can feel too ashamed of their bodies to get regular medical care, meaning that any potential problems may not be identified early enough for successful treatment. Yet, while it is true that obesity carries certain risks, it also is the case that the ideal weights used by insurance companies to indicate health risks change periodically, and may be misestimated. Moreover, as we are coming to know, some people simply are heavier than—and just as healthy as—those whose weights meet the cultural ideal.

Body Decoration

Beyond size and associated body shape, for which each culture has slightly (albeit sometimes radically) different ideals, cultures have ideals for how and when bodies are cleaned, dressed, and decorated. They also have ideas about which body parts must be manipulated. Sometimes, this manipulation involves the permanent physical modification of the parts in question. We learned in Chapter 11 about rites of passage that create men and sometimes women through ordeals such as circumcision, which leave a permanent mark. Other modifications are made for the sake of beauty ideals. Examples are provided below as we discuss first the various dimensions along which body decoration and modification vary, and then the different functions they serve.

Dimensions of Variation

Body decoration and modification may vary along a number of not mutually-exclusive dimensions. For instance, they may be: (1) Voluntary or imposed (stigma); (2) Affiliative (normative) or disaffiliative (deviant); (3) Flexible (temporary) or fixed (permanent).

Voluntary or Not

To begin, body decoration or modification can be voluntary or imposed. For example, while many people in the United States today elect to tattoo themselves, in other times and places tattoos were reserved to mark outcasts, who were forced to submit to them.

It is well known, for example, that the Nazis tattooed their prisoners. They stigmatized many groups of people this way, not so much for breaking behavioral rules (although there was some of that), but for being of a certain heritage or biological lineage (the Nazis subscribed, mistakenly of course, to an extreme and reductionist form of biological determinism). A catalog of stigmata was used by the Nazis during the Holocaust to mark individuals as specific kinds of outsiders to the so-called Aryan race. Conversely, the blonde, blue-eyed, Christian, heterosexual body, robust and with all its normal parts, conformed to and supported Nazi social norms and expectations. Accordingly, people who would affiliate with the Nazi community voluntarily strove for such a look. Self-elected participation in physical fitness training camps, for example, provided one avenue to achieve that outcome.

Affiliative or Not

Another dimension of variation in body decoration or modification is whether or not it is affiliative or disaffiliative. If **affiliative**, it will symbolize one's affiliation or identification with a particular culture or subculture. The military uniform worn properly, for instance, tells others that its wearer is a team player, willing to conform, and interested in supporting his or her country.

A uniform—or other forms of body decoration or modification for that matter—may be 'disaffiliative,' too, signaling or placing distance between an individual and a particular group. A disaffiliative decoration or modification marks an individual as deviant, or as undesirous of affiliation with or of signaling membership with the norm or with the mainstream. The punk hairdo carried this meaning in the late 1970s. Tattoos, until recently, did too.

Here I should note that affiliative body modifications and decorations often entail an attempt to 'humanize' the body so that, in addition to affiliating with a particular group, we are, in conforming, affiliating with or demonstrating our very humanity. Conversely, to leave one's body in a 'natural' state, with hair uncut and unwashed, for instance, or to remain undressed is in effect disaffiliative. Many cultural myths regarding outsiders depict them in such a fashion, as wild women or men. Accordingly, in joining a group (or rejoining one as, for example, in the case of a man or woman who has been away at war), a person often will undergo some modification practices, such as hair cutting. In some contexts people must wash their faces and hands, and sometimes their feet as well. In mainstream U.S. culture, people who have been away from home sometimes become very preoccupied with a felt need to shower and shave, so as to 'feel human again.' We manipulate our bodies in ways meant not only to affiliate with our group or with humans, but also to disaffiliate with animals or outsiders.

We also do it for gender-affiliative purposes; we modify and decorate ourselves to feminize or masculinize our bodies. Mainstream U.S. women, for example, shave their legs and armpits. Feminine shaving is not a hygiene practice, although we will sometimes rationalize it as such. While other societies do not think about and sometimes even favor hair in those areas, feminine shaving is, in Western society, a means by which to differentiate men and women. Some theorists would go so far as to say that, in shaving her body, a woman seeks to turn herself back into a child (children do not have body hair), symbolically reinforcing the patriarchal social order through her doubled removal from power. Whether or not that is so, feminine shaving definitely places a claim on a gender affiliation while disaffiliating a person from the masculine.

Here, let us bring in the Beatles, who ruled rock and roll in the 1960s (at a minimum, some would say). Even in their early years the Beatles made a certain statement with their chosen mode of body decoration and modification. That they did this was not wholly up to them; we all are available to other people freely as texts to be read, or read into. People do look; and people do talk.

The Beatles' hair was one of the main things people looked at and talked about (see Figure 12.3). Even at first, it was a little bit longer than the norm. Some people did see this as somewhat radical, yet because the

Figure 12.3 The Beatles as they arrive in New York City in 1964. *United Press International.*

Figure 12.4 George Harrison in the Oval Office during the Ford administration, December 13, 1974. *Photo by David Hume Kennerly. White House photo courtesy of Gerald R. Ford Library.*

Beatles performed in suits and with shaved faces most people could live with and indeed looked past it.

In contrast, any picture of the Beatles in their later years shows a real difference in grooming, hair length, and dress. Here, there is no question that their body decoration and modification choices reflected disaffiliation from the mainstream, and the so-called older generation. Disaffiliation was embodied in the ways that these musicians carried themselves, groomed themselves, dressed themselves, and marked themselves physically (see Figure 12.4).

Disaffiliation went beyond the body surface, too; the type of labor they did with their bodies, as musicians, and, moreover, their physical lifestyles in no way conformed to the type of labor or lifestyle that would have been considered respectable for young men in those days. Some in the mainstream were actually quite frightened by the Beatles' disaffiliative use of their bodies. They felt the same way, for similar reasons, about musicians with dreadlocks, who wore leather, or who appeared as punks. Yet disaffiliation and rejection of the mainstream was not the only message that such musicians sent with their bodies. They also made an affiliative statement in regard to others like them; the Beatles, for instance, wished to be part of the pro-peace faction in the '60s revolution.

That the Beatles' hair did change brings us to a third area of variation: the fixity or flexibility of a body decoration or modification form. Hair can be cut, dyed, braided, beaded; and it grows—changes to hair are generally flexible or impermanent. Flexible decorations and modifications are those we can take on and off much like a judge or barrister in a Commonwealth nation puts on and takes off the white wig that sets his or her role as a judge. Once the judge takes off that wig—and the robes and so on—other roles come to the forefront.

Permanent or Not

Importantly, the wig and robes, like other work uniforms, support the embodiment and enactment of an entire identity that is nevertheless partial or flexible. What we wear—our social skin, so to speak—does have a very real impact on how we comport ourselves. When someone puts on a police officer's uniform, he or she will feel like a police officer and therefore act like a police officer; that's as it should be.

Tattoos, in contrast, are generally fixed or permanent markers—particularly in times and places without access to removal services. Perhaps because those now exist, a better example of a fixed or permanent body modification might be teeth filing in Sumatra. There, people file their teeth to points as a way to create and maintain themselves as human. Other examples of fixed decorations or modifications include the use of metal neck rings or coils (which actually do not so much stretch the neck as push down the clavicles or collar bones) among the Kayan Lahwi, now of Burma; lip-stretching plates or plugs worn among, for example, Mursi and Surma or Suri women in Ethiopia and Suyá men in Brazil; ear lobe plugs; and scar patterns carved into the flesh. In contrast, face, body, and nail paint provides a temporary form of decoration, as does shaving your head.

Returning to tattoos, which have, after all, a very long history as a fixed form of body art, let us consider the Māori situation. These tattoos stand in contrast to decorations and modifications that are flexible or partial—that can be taken off or put on as one likes; or as is appropriate according to what one does for a living; or in keeping with cultural expectations as, say, for certain ritual events. Among the Māori, who are native to New Zealand, high status men and women traditionally are marked as high status with facial tattoos. Another Māori person can tell from both the extensivity and shape of someone's tattoos what his or her rank is. From someone's tattoos, one knows how to behave in his or her presence—and how he or she will behave toward oneself. This is simply by reading the marks on the tattooed person's face. The Māori also have a tradition of extensive body tattoos and these too form part of the text. Unlike in the present-day United States, the designs are chosen not for personal reasons but rather for what they connote in the Māori tradition regarding a person's social status, kinship ties, and accomplishments. They correspond, then, to things of social and not necessarily personal importance.

In some populations, rather than to tattoo such information onto the body, scarification techniques are used. In scarification, scars are made on the body purposefully, generally in geometric patterns. Small cuts are made over and over to generate particular designs, which cover broad swaths of the body, such as the belly, back, or forehead. The cuts are made with a clean and very sharp tool, ideally. They often are washed and covered initially with some kind of medicinal ointment or poultice to promote good healing (all cultures have some form of pre- and post-treatment process in place to insure against infection; a poorly healed scarification job is not only a health hazard but also a hindrance to the communications the scars are meant to send, so it must be protected against). Some Australian Aboriginal people have practiced scarification; many African groups do so. Scarification seems to work better on darker skin than tattoos for communicating vital information such as what tribe one belongs to, how many children one has had, whether one is married, or of high status or a chief. It also does something that tattoos cannot generally do: the raised lines or shapes made in the scarring process can be felt. There is a decidedly tactile aspect to this form of body modification.

The contrast between whether or not a given group is prone to fixed or flexible forms of body decoration or modification might seem, on the surface, like a contrast between modern or contemporary peoples and traditional peoples. It is true that they seem to be distributed that way, with more flexibility among the former and more fixed forms among the latter. Yet it would be much more accurate to say that the contrast is between closed societies on the one hand and open societies—or really open class societies—on the other (see Table 12.1). **Closed societies** are those in which a person's social status is ascribed; often it is based on birth or anchored in a rite of passage from which there is no going back. **Open societies** are those with inbuilt class mobility (or at least the myth of such). Closed societies favor or have more forms of fixed body decoration and modification, which makes sense because little that is expressed in them will change; open societies have more flexible forms. This reflects the plasticity of one's social status in an open society, and the fact that one may go in and out of many social roles over even a short course of time there.

Table 12.1 Society Type and Permanence Pattern or Typical Body Decoration/Modification Form

	Fixed form (permanent; e.g., patterned scarring)	Flexible form (temporary; e.g., team uniform)
Society Type	Closed	Open
Source of Social Status	Ascribed	Achieved

Functions of Body Decoration and Modification

In addition to the three dimensions of variation for body decoration and modification (voluntary-imposed, affiliative-disaffiliatvie, fixed-flexible), we can consider its three functions: to (1) communicate, (2) protect, and (3) transform.

Communicate

I actually began this chapter talking about the first function; I noted that the markings or decorations on or modifications to a person's body—indeed its very shape—can communicate to other people who that person is and what she or he stands for. One thing the marks might communicate is that a person has adopted the correct attitude, and supports the cultural values of the group. Adopting the same aesthetic or beauty ideals that your group holds communicates this message of affiliation.

However, more than simply being affiliative, modifications such as tattoos or scarification and decorations such as medals or maybe a crown, can also communicate to other people important facts about, for instance, your social position (who but the king or queen wears a crown?). Such symbols even can communicate information about your achievements. If a person wears medals, for example, from fighting or otherwise in World

War II, the person's peers and compatriots will know that much more about the person without ever having verbally been told. Likewise, some tattoos mark achievements. All these are examples of the communicative function of body decoration and modification.

Protect

The protective function leverages the meaning response (described in Chapter 10) to provide a sense of safety. A symbol, embedded or emblazoned upon your body, evokes a meaning response in you. If you know that a particular shape or symbol is meant to protect, you will feel protected and go about your activities without being nervous and therefore not making nerves-related mistakes. This is one reason for wearing amulets or even particular perfumes or tattoos.

Clinton Sanders (1988) tells of a man who wanted a big, brutal bee tattoo; he thought it might scare away real bees, to which he was deathly allergic. If he encountered real bees when under the protection of a mean bee tattoo, he would feel safe and calm rather than to experience an intense stress response. If he lived where there were many bees, this would surely cut back on his risk for chronic stress related illness. The protective function of body decoration or modification can be directly physiological.

Transform

A third function of body decoration and modification is to transform a person from one status to another. We did discuss the transformative function already, albeit indirectly, when discussing rites of passage (Chapter 11). Undertaking a permanent form of body modification can be a very important **performative** part of a rite of passage, so it not only serves later to communicate that the person has been initiated into, say, manhood, but actually helps to create this transformation. For instance, the pain from a particular operation or simply the anticipation of it may create in the initiate an altered state or a sense of intensity, making the rite more meaningful and the lessons imparted more memorable. Further, through enduring it someone may prove himself a man or woman. The transformative function of body decoration or modification occurs in the liminal period of the rite.

Keeping a Relativistic Perspective

Sometimes, when outsiders consider the pain and so forth of a certain body modification process—particularly one not generally done in their culture and so not seen as crucial, or one done with children—they reject it out of hand. This begs the question of its culture-specific utility, and necessity. The importance of actually undergoing a body-modifying rite of passage is paramount for people if it is normative in their society. Remember, a person without the proper markings has not identified him or herself as a member of the group. To fail to conform in that way would be to cut oneself off from the possibility of, for example, getting married and achieving other life goals held dear by one's culture. The stress generated this way would be high indeed.

Moreover, if outsiders turn the mirror on themselves, they may see that their culture, too, entails practices that can appear brutal. Take, for instance, orthodontia in the United States; this long term dental modification technique can be quite painful, physically and socially. While sometimes done for clinically legitimate reasons, it generally has cosmetic goals. Outsiders might see orthodontia as both vain and, when imposed on children, barbaric. They may be especially primed to see it as abusive in light of their knowledge of other mainstream U.S. bodily practices imposed upon children. For instance, forced early weaning and forced solo sleeping for young infants are seen not only as sadistic and inhumane but also as incomprehensible in some cultural worlds: how could a parent be so cruel to her or his own child?

Appearances to others aside, we, too, are expected to present to the world a face of stoic forbearance during certain painful activities; we, too, have practices where enduring the pain is part of a transformational process. Take, for instance, U.S. practices surrounding tattoos, face lifts, high-heel shoes that pinch, pectoral implants, vasectomies, wax-based hair removal, and so on. Even when undertaking fitness-focused exercise we often say "no pain, no gain." We have hopes and wishes for ourselves and our children just as do cultures that use body modification techniques in formalized and full-blown rites of passage. In sum, although there is cultural variation in how diverse populations will practice body decoration and modification and body shaping activities, and there are concomitant variations in bodily outcomes (including for health), we all do practice them. We do so with what are, ultimately, very similar and very human goals.

13

Kinship: So Relative

This chapter prepares you to:

- Describe the three most basic kinds of kinship, paying particular attention to cross-culturally variant dimensions of consubstantiality

- Explore diverse understandings of when, how, and if life begins and explain the consequences that various understandings have for practices such as abortion

- Explain how cultural ideas about kinship underwrite cross-cultural diversity in family-building strategies

Kinship is a system of embodied social relations. While in the West kinship often is treated as a biological fact, it is not so clear cut or simple as that. In this chapter, we examine the bioculturally diverse basis—or bases—for kinship in various cultural contexts, and we explore the ramifications of these for family building strategies.

Kin Terms and Lineages

Kinship terms name various kinds of kin. Although one might think that the explanation stops there, different cultures have different kin terms—and so different kin roles. For example, some cultures do not distinguish at all between siblings by gender; we do: we have 'brothers' and 'sisters.' Other cultures include information about relative age in their terms for siblings, so one could have an 'older sister' as well as a 'younger sister'—and 'older' and 'younger' brothers, too. Notice that I have to resort to compound words in English; in other languages, specific, discrete words can exist; for instance, in Japanese one's own younger sister is *imouto* while one's older sister is *ane*.

Other cultures also may be more specific than our own when it comes to cousins. For example, if a culture prefers that cross-cousins marry, there will be words to distinguish them from parallel cousins (cross cousins are born to a parent's opposite-sexed sibling, such as one's mother's brother's child; parallel cousins are born to a parent's same-sexed sibling, such as one's mother's sister's child).

Although kin terms, and so kin roles, vary across cultures, what kin terms accomplish does not. What they do is to map out the rights and obligations that people have toward one another as kin. That explains why, in some contexts, it is crucial to know who your parallel or cross cousins are (you might have to marry one of them). It is worth noting also that the roles specified for given labels (for example, father, mother, daughter, son) vary in different contexts. In the early days of the Israeli Kibbutz movement, for instance, the role of mother or father did not include an obligation for childrearing; children were raised communally by kibbutz members who specialized in the childcare role (see Figure 13.1).

We often use kin terms to call out the rights and obligations they index. For example, in mainstream American culture, when a parent sees his or her daughter or son leaving the house dressed in a way that contradicts parental common sense, he or she may say "No son [or: daughter] of mine is leaving the house dressed like a [fill in the blank]." In doing so, the parent is calling on the son's or daughter's kin-based obligation to honor or at least not to shame the parent and the family as well. Likewise, when a parent needs help with the kids, and so calls out to his or her partner "Mother [or: Father], could you please come put your children to bed" (or

Figure 13.1 Communal child rearing was common in the Israeli Kibbut movement's earlier years. *Kibutz Einat Archive via the PikiWiki—Israel free image collection project.*

give them a bath, or discipline them), the verbalizing parent is seeking to mobilize in the other the kinship obligations that go along with co-parenting. Kin terms are used to remind people of their responsibilities as members of the family, as well as to remind them of the right others have to expect certain kinds of behavior from them.

Kin terms also map relatedness. When we hear somebody say 'uncle' or 'father,' we know how they are related—at least if we know their culture. If someone in mainstream U.S. culture is related to a person as his or her father, he will have a very different kind of relationship to that person than he would if he was that person's uncle. However, relatedness is culture-dependent. While in most Western cultures the father and son are considered more closely related, in some cultures it is the nephew and uncle who are the closer kin. This has to do with what kind of **lineage** a given culture prioritizes as 'real' or binding. In other words, it has to do with how they track heredity.

Anthropologists have identified many versions of lineality, or many ways to figure the family line. Cross-culturally, we have **matrilineages**, in which heredity is figured through the female line. We have **patrilineages**, in which heredity comes down through the male line. We also have bilateral lineages, in which heredity is tracked through both women and men (this is how many people in the United States figure heredity). There are other variations as well.

In a patrilineage, children represent an extension of the father's lineage, and they generally take his name to show this. In a matrilineage, the same is not true. Children in matrilineal societies are part of their mothers' kin lines. Thus, while a father in a matrilineal society (let's call him Jack) may love his children dearly, and care for them as obligated to through the bonds of marriage and fatherhood, his real interest—kinwise—is in his sister's offspring. Jack's sister's children are the next generation of heirs for his lineage, and so he has great stakes in doing well by them. The relationship he has with them is termed an **avunculate** (the term uncle is embedded in there, making this an easy term to remember).

The avunculate really is a wonderful invention. While a mother who gives birth to her children is concretely assured that she had some part in their creation, a father does not have such concrete proof of paternity. Short of inventing a way to claim it, as through naming and familial authority conventions such as we find in patriarchal societies today, a man may be better off focusing his energy on children he *knows* are in his lineage: his sister's offspring.

The avunculate is a very important relationship in many matrilineal cultures. When Jack's nephew marries, for instance, the boy and his new wife may assume avunculocal residence (they will come and live with Uncle Jack). This living arrangement is in contrast to matrilocal residence (when the newlywed couple is meant to reside with the wife's kin), patrilocal residence (when they live with the husband's father), and neolocal residence (when they form a new household entirely).

Who Is Related, and How?

Kinship is about connectedness—embodied connectedness. It has a cultural aspect, sure; but it also is biological, and so exploring kinship adds to our knowledge of biocultural diversity.

In the past, anthropologists mapped out lineages using certain disciplinary conventions such as circles for women, triangles for men, equal signs for marriage, vertical lines for parent-child relations, and horizontal lines for sibling ties (see Figure 13.2). Those conventions are about 100 years old, and they reflect the family structure idealized by the scholars who invented them. The conventions are thus well-suited to depicting a nuclear family, with one mother, one father, and some descendents.

That said, they do not work very well when applied cross-culturally. It is hard, for instance, to demonstrate the avunculate with them (see Figure 13.3). They do not even work very well to describe the reality as it

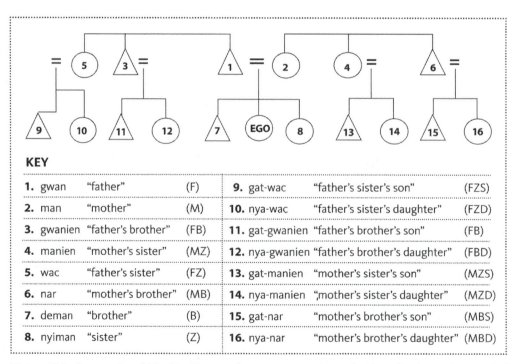

KEY

1. gwan	"father"	(F)	**9.** gat-wac	"father's sister's son"	(FZS)
2. man	"mother"	(M)	**10.** nya-wac	"father's sister's daughter"	(FZD)
3. gwanien	"father's brother"	(FB)	**11.** gat-gwanien	"father's brother's son"	(FB)
4. manien	"mother's sister"	(MZ)	**12.** nya-gwanien	"father's brother's daughter"	(FBD)
5. wac	"father's sister"	(FZ)	**13.** gat-manien	"mother's sister's son"	(MZS)
6. nar	"mother's brother"	(MB)	**14.** nya-manien	"mother's sister's daughter"	(MZD)
7. deman	"brother"	(B)	**15.** gat-nar	"mother's brother's son"	(MBS)
8. nyiman	"sister"	(Z)	**16.** nya-nar	"mother's brother's daughter"	(MBD)

Figure 13.2 Conventional kinship diagram for two generations of a Nuer (Sudanese) extended family; Nuer kin terms also are depicted. *Courtesy William Balée,* Inside Cultures, *2011.*

was on the ground 100 years ago, let alone today. One has to add arrows, connecting lines, comments, and so forth to depict strong extra-nuclear relationships as in the avunculate. There is no inbuilt way to distinguish siblings by age, although sibling gender can be depicted—so long as it is stable. What if it changes over time? How might we depict a son that once was a daughter, or divorce and remarriage? There is no easy way. The conventions assume that gender, and marriage, will last forever.

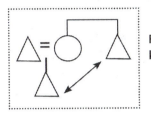

Figure 13.3 The avunculate is difficult to map using conventional kinship diagram symbols. *Courtesy William Balée,* Inside Cultures, *2011.*

Beyond not being able to capture or accommodate certain cultural realities, use of the kinship symbolism by anthropologists to describe lineages in other cultures resulted, in effect, in forcing square pegs into round holes. It described (or tried to describe) families with the anthropologist's kinship categories and from the anthropologist's point of view—not the native's. It certainly is important to describe kinship systems and to chart genealogies; often, information about who is related to whom, and how, is quite enough to unlock the reasons for particular behavioral patterns. In many cultures, nearly all behavior is guided by kinship rights and obligations. However, if kinship descriptions are ethnocentric to start with, crucial information regarding them will be missing from the ethnographic record. A swath of biocultural diversity will remain unaccounted for.

The question of viewpoint has been explored in relation to the experiences of immigrants to the United States who would bring their kin here to live with them. In the United States, courts of law depend on DNA evidence to define kin relationships. However, the DNA model does not match the kinship model used in many immigrant families. DNA samples may indicate that sons and daughters are in fact not 'related,' and so they may be denied entry. The tragedy this represents for many families cannot be overstated. It affects people with differing conceptions of kinship than those held by U.S. authorities; for example, if Uncle Jack tried to bring his nephew in, he would be denied—although he might be offered entry for his children-by-marriage even though to him they are nowhere near as related. (Their mother, of course, would not be happy to part with them as they would be her lineal kin.) Another problem with insistence on DNA testing is that it may bring to light extramarital affairs that had long been forgotten or which were outweighed by the parent-child relationship that mutual love and caring can create (Swarns 2007).

Now one might say that DNA is the only true indicator of heredity, which derives from shared genes. That is a scientific fact: it is not in question. What is in question here is not the validity of genetic kinship but rather whether other kinds of kinship exist—and what kinds count. What kind of relatedness is important in or to a family? This can be variously defined depending on what culture we are talking about. It certainly varied, in reality, for many of those old scholars who included in their work the kind of kinship charts described above. Even those who live in mainstream Western society today and who have taken biology and genetics classes may practice kinship in very different terms than those provided in the genetic model of relatedness.

As recent scientific discoveries have shown us, parents' or caregivers' influence even on children who are not related genetically can be vast. That is, children can, so to speak, inherit traits even from non-genetic kin. This is due to the powerful developmental and epigenetic imprints that can be made on children by the environments in which they grow—by the environments that caregivers and parents, and whole cultures, create.

Kinship Types

The kinds of relatedness that social scientists have seen when looking around the world at different cultures, and when looking back through history to different times, boil down to three types: (1) classificatory, (2) affinal, and (3) consubstantial. The first category is in some ways a non-starter because **classificatory** kin are people who really are not kin but who are treated as such. These kinds of so-called kin also have been termed 'fictive,' but I prefer 'classificatory' as it lacks the negative connation that 'fictive' had when it first entered the anthropological lexicon. It was introduced along with those early kin charts and applied, sometimes quite callously, to kin relations that, in Western eyes, were fictional. Today, many of those relations are recognized as actually being consubstantial, but I will get to that later. As for fictive or classificatory kinship today, we use it to refer specifically to cases in which people knowingly use kin terms to refer to people who are not kin. A good example of this is when Grandma introduces her best friend to you as her sister or your aunt. I had an Auntie Doris; she was my granny's best friend. The two were not related, but they treated each other—relied on each other—as if kin. So Auntie Doris was a classificatory aunt of mine. Another example of classificatory kinship is when someone says that a person is like a daughter or son. Applying a kin term like this is tantamount to listing out the set of rights and obligations incumbent on that class of kin. A friend announcing, "You are like a sister [brother] to me," accomplishes the same.

Affinal kin also are not in one's lineage, but they are real kin because they have been created as such by the law. In other words, and in Western terms, while we share no DNA with our affinal kin, we are still kin because we have entered into a contract that makes it so. Husbands and wives are affinal kin. Adopted sons or daughters are, too. Spouses and adopted children are treated by law just like other lineage-linked kin; in fact, in present day United States they sometimes actually have more legal rights and responsibilities. Your spouse, if you have one, has more say over your medical care, for example, than does, for instance, your sister, or your mother for that matter (if your spouse is alive and you are of legal age, that is).

Last but, as we shall see, not least, are **consubstantial** kin: people who are related to us because we share some substance in common with them. We are conjoined by the substance we share. These kin generally belong to one's lineage or, as we used to say before the discovery of DNA, one's bloodline—and indeed, these kin used to be called **consanguineal**, in reference to that shared blood. 'Consanguineal' still is used by some, but it is more correct to treat it as the subcategory it is. 'Consubstantial' is the better word choice for the scholar, because this category of kin is not limited to those thought to share blood or even its Western replacement substance, DNA. There are many substances that, when shared, create the bodily bond—and experience—of kinship.

Forms of Consubstantiality

Blood Kin

The idea that all 'real' kin are consanguineal comes from the folk or popular logic espoused in many if not all cultures that babies are somehow grown from the blood of their parents. In some cases, the word 'blood' is generically used to mean any number of bodily fluids; it can be a term, then, somewhat like 'humors,' as in the humoral system of the ancient Greeks.

Jamaicans distinguish between two kinds of blood: red and white. Red blood is the same kind pumped by the heart; white blood includes semen, breast milk, and other light-colored fluids, including female sexual or reproductive discharge. Once we understand blood broadly, we can see how some cultures saw blood kinship as due to the mixing of male semen and female body fluids. While some cultures hold that a fetus, in the womb, takes in parental blood substances through its mouth, others teach that blood substances simply accrete or build up in layers once the fetus has been created (sometimes conception happens when parental bloods mingle). Until recently, Anglo-European culture shared this kind of belief.

The power of various bodily fluids to relate people can be used in love magic as well as in kinship creation. Love and kinship actually may be thought of as one and the same: both entail enduring solidarity and mutual support. In some cultures, menstrual blood can be used to create a bond between two (or more) people. A menstruating woman might, for example, put a little menstrual blood into some food, perhaps soup, or some coffee; she then serves it to the person she would like to tie or bond to her—maybe a man to whom she would like to be married, or her husband if she has one and he has strayed. Men, of course, also have substances with which they can do such magic. In incorporating some of someone's substance into your body, a relationship is created. Sharing blood has, historically, and generally speaking, provided the majority of cultures a key trope for this.

Genetic Kin

Nowadays in the West, the blood that begets kinship has been replaced in the popular mind with genetic material. Anyone who has passed a high-school biology class knows that **gametes** (eggs and sperm) contain chromosomes and that those chromosomes are what link a child's physical body back to its parents.

The idea of shared genes is in fact such a key part of the mainstream idea of kinship in the West, and kinship is still so important to us, that we often bring genes up when noting a family resemblance. In fact, we often search for such family resemblances, as if to prove to ourselves that there is indeed a genetic link in a given family (Becker, Butler et al. 2005). The mainstream idea that children should be borne in wedlock and other assumptions regarding the ideal family unit come into play in such searches. If one child in a family has, say, brown hair, in a family where everybody else's hair is blonde, people look for a genetic explanation. One helpful relative may offer, "Well, it was probably from your great grandmother. Did you know she had brown hair?" Another may confirm it. Then another, who has not heard the first conversation, might offer that grandpa's brother (on the other side) was a brunet. Whichever story the family agrees upon in effect does not matter; the important thing is that a connection can be claimed. If not, then the child's position within the family might be called to question, at least according to longstanding ideals regarding American family structure and the place of (genetic) kinship in it.

Indeed, children's parents may participate in such family history-making even if, or in fact because, they know differently. They may know, for example, that the child was adopted or otherwise does not share genetic material with the side of the family that relatives are pointing to. They may remain silent about this due to understandings regarding the importance of genetic connection. In this way, they become co-conspirators in supporting mainsream kinship and family building ideals that actually are in some sense oppressive to them.

Nurturance Kin

In other cultures, and in some subcultures in the United States, there are other options for kinship creation or confirmation besides those offered by genes. In many cultures, the main substance that connects people, creating kinship, is love or nurturance. This entails the effort put into raising a child, and the emotions invested. Feeding a child, providing guidance, paying school fees, creating a loving home—all this is enough, in some contexts, to create an actual kinship link with that child. The love and nurturance given is substantive; it creates a consubstantial link. It may not have a material presence, and so may seem non-substantial to outsiders, but like electricity or sound waves it is indeed there; it has real, substantive, measurable effects and is considered as if embodied by insiders.

Milk Kin

Breast milk is often considered a valuable substance for solidifying kin ties that might already exist or for creating kin ties where there were none to begin with. The Azawagh Arabs, who we met in the last chapter, practice milk-based or lactational kinship (Popenoe 2004). Here's how it works. Under a number of conditions, a woman may nurse children who are not biologically hers. First, an already lactating women might help a new mother whose own milk has not come in (it does take a few days for the body to begin full-scale milk production) or who, possibly for health reasons, has an insufficient milk supply. Second, she might help out if the biological mother has to be away from the baby and the baby gets hungry during that time. Third, she may do it intentionally to create a milk tie.

Why would a woman seek to do that? She might see a benefit in manipulating her children's possibilities of marrying certain people. Thus, she might strategically breastfeed a little girl she does not want her son to marry. The significance of a milk tie does not stop in this generation. While the first generation of children related by milk are related as siblings, the offspring of their marriages (made with non-sibling partners, of course) will be cousins—and cousins are preferred as marriage partners (this is part of the 'containment' approach that this culture subscribes to;

see Chapter 12). So a breastfeeding session with the child next door might in fact be part of a long term plan for one's grandchildren's marriages.

Moreover, and perhaps more importantly, milk kin are kin just like blood kin, and so must welcome each other into their homes as family. The more kin a person can claim, the larger his or her social support network will be. This is one key reason for creating milk kin.

Commensality

The Importance of Co-ingestion

The baby that becomes milk kin does so by ingesting the bodily product of its milk mother. Ingestion also is a key means of other forms of consubstantiality.

Commensality—sharing a meal or breaking bread together—signifies and creates and helps maintain relationships. This is one of the reasons why rituals that reinforce group solidarity often involve food sharing. Commonly, the food shared is a staple item in that culture—something that members feel they cannot live without. For example, in Japan, there are many rites involving sharing rice. In a traditional Japanese wedding ceremony, sharing a bowl of rice becomes crucial to solidifying and demonstrating the kinship bond being created by husband and wife, or between the husband's family and the wife's family.

In other weddings, such as Macedonian Orthodox ones, bread is shared. In Macedonian tradition, the bread should be sweet, so that the bride and groom will share a sweet life together. Besides the couple, their families all eat some of the bread. Thereby, these two separate families and two separate individuals join together. The social distance between them has been eclipsed by the fact that they all, within their bodies, carry a piece of the same shared substance. They are now, in effect, one people. The same symbolism holds for shared wedding cakes.

Without food sharing, many of our social relations would lack glue. Just as crucial, however, are rules about with whom food should *not* be shared. Sharing food with them would create relationships that are too intimate to be acceptable—relationships that are taboo.

The Hua

Like all people, the Hua of Papua New Guinea have food sharing rules; these are described by Anna Meigs (1997), whose work we explore here. Among the Hua, some rules—absolute rules—have to do with essential or absolute properties or qualities in given foods and the understanding that eating them would recreate those properties within the eater, via contagion. Women, for instance, do not eat men's foods; and vice versa.

We do the same, at least minimally; consider for instance the gendered patterns of what men and women drink at bars, and how bartenders decorate or in what kind of glasses or containers they place 'lady' and 'guy' drinks when they serve them. (Consider also how eating certain 'ethnic' foods—hummus, posole, kimchi, Marmite, grits—can help members of an ethnic group to reinforce that ethnic aspect of their identity.)

Other rules have to do not with what foods to eat or not, but from whom to accept foods. These rules focus on the essential properties or qualities of the people who offer the food. Meigs calls these rules relative rules, and while the other kind of rules (absolute rules) are about contagion, relative rules are, in a sense, about cannibalism—or really, how to avoid it. The idea behind relative rules is that whatever a person cooks or produces in his or her garden thereby contains some of his or her *nu* or vital essence.

According to the Hua, nu inhabits all of our body parts, including our flesh and blood as well as things extruded, such as breath, hair, nails, skin flakes, pus, feces, urine, and reproductive tract fluids. Anything that has come in contact with something from one's body, then, will absorb some of one's nu. If your sweat gets onto another person's skin, it can move through the pores into his or her body, thereby creating a kinship connection, or potentially doing so anyhow.

When eating food, one also is eating the nu of the food preparer and sometimes even of the food producer. When a Hua person works in his or her garden he or she may breathe on the plants. Sweat from his or her body goes into growing them. A fingernail may break and fall into the soil that nurtures the gardened items. Bits and pieces of a person therefore literally enter the food he or she has produced. The same thing happens to us when preparing food—when making a sandwich or baking cookies or roasting turnips. Our hands touch that food and, as they do, a little bit of our essence is transferred. A little bit of our breath, for example, falls on the cookie batter as we mix it. A hair might fall onto a sandwich or some skin flakes may come off onto the bread. When someone else eats that food, that someone else eats a little bit of us, too.

The Hua find more significance in this than we do, and so, for example, Hua parents should not eat food from their own children: to take and ingest food that one's own child has prepared or produced is almost like eating that child for dinner. Likewise, people should not eat their own best produce because that would be like eating themselves. Food sharing or commensality among the Hua creates consubstantiality. It is not the food matter itself that creates this. It is the vital essence that the food carries: its preparer or producer's nu (Meigs 1987).

Importantly, nu conveys not only the substance of kinship but also feelings. Nu from an enemy thus can be destructive and must be avoided, unlike the nu of family members who, in theory, wish to support their kin. Many Westerners subscribe to some version of this theory, eating freely from kin but maintaining a wariness of food offered by strangers and surely refusing food prepared by people with moral failings, no matter how well their hands are washed or kitchens cleaned.

Building up kinship among the Hua takes time. People interested in it, therefore, purposefully add bits of their bodies to others' food to speed up the process. Mothers solidify ties with their children through breast milk as well as by rubbing babies with their own sweat and body oils. Parents cut their skin and feed the let blood to their children. In the past, sometimes even the flesh of deceased individuals was fed to others to ensure lineal ties.

This method for creating relatedness makes affinal or legal kinship irrelevant. Spouses may start out as mere affines but become conjoined consubstantially after a time. It does take a while—maybe fifteen years—because to marry, a man and woman must not be directly related, meaning there is no notable baseline of kinship to begin with for them. Still, after enough time, and enough consubstantial exchange, spouses are linked bodily. The family will hold a rite of passage in which a woman gives her husband, in public, some leafy greens from her garden (in Hua culture, greens are loaded with their producer's nu). The husband's elder brother proclaims that the husband can eat freely from his wife now—as freely as from his own mother (these are the terms expressed). The man and now, too, his family (which until then could not eat at all from the wife because she was unrelated in their system) can share food. The wife is thereby transitioned from a mere in-law (affine) to a consubstantial relative.

When Does Kinship Begin?

Another interesting thing about the Hua—and there are many similar-thinking groups—is that the currency of nu exchange matters: real kin are people with whom you are right now exchanging nu (Meigs 1987). That is, Hua kinship does not last forever. In the West, most hold that lineal kin always are kin. Parent-child links cannot be denied (although they may be ignored). Amongst the Hua, however, unless people are constantly sharing substances, filling each other up bodily with each other's nu, a kinship bond can dissipate; it can fade away.

This is one reason why people from certain groups do not like to go away to college or to migrate for work far away from their kin. By doing so—unless they come back periodically for a reinjection of kinship-building substances—they may cease to be family members.

The larger question this brings up concerns when kinship bonds can be established. In Western, scientific thought they are established prenatally or before birth, through conception, before a child is born. In this, kinship is a static thing. Kinship is a closed system, and immutable or impervious to change. Once sperm and egg have come together, an individual's relationship to his or her mother and father has been created and sealed. It cannot be changed.

Instead of having a static, closed, prenatal kinship system, the Hua—and many others, too—have a dynamic, open system. Kinship can be changed. Open systems like this can accommodate postnatal kinship creation (after birth) because kinship, rather than being instantly created, is built up over time by accretion. Indeed, in some cultures, even prenatal kinship depends on that; continued sexual intercourse is necessary to feed and grow the fetus. Every act of intercourse adds substance to the growing clump of matter that over time becomes a baby. In this kind of additive as opposed to instant system, a baby can easily have more than one father (see Table 13.1).

The flexibility of open and additive systems increases one's kinship options, for instance in terms of having multiple parents. However, the demand for continued maintenance that this kind of kinship generally calls for has costs, too, one of which is in the sheer effort of upkeep. It also has ramifications to how adoption is experienced.

A person might have come from a particular parent's body, but when that parent stops doing what parents need to do to solidify parent-child kinship, his or her child can become the son or daughter of another parent

Table 13.1 Features of Prenatal and Postnatal Kinship

	Prenatal or Birth Kinship	Postnatal Kinship*
Onset	Instantly given (e.g., when egg and sperm meet)	Additively achieved (e.g., with breast milk)
Flexibility	Static—closed to change	Dynamic—open to change
Sustainability	Endures under all conditions once created	Needs boosters for long-term endurance

*Note that in societies where postnatal kinship creation is possible, prenatal ties also can be made, generally by similar means (e.g., accretion).

(or parents) completely. An adopted child among the Hua, for example, has no claim on a genetic parent simply because that parent is genetically related. Parents give up that title and any related claim on a child when they fail to share nu. In such a context, any feelings of linkage are culturally minimized; they have no cultural basis. Conversely, the profound sense of kinship often expressed by U.S. adoptees toward their genetic parents and vice versa—even without actually knowing them—is catalyzed by U.S. cultural values.

When Does Life Begin?

Related to the question of when kinship starts and how it is established is the question of when each individual's human life begins. The way that a given culture answers this query—or whether it even asks it—has ramifications for decisions that are both biological and cultural in nature and which, in turn, have concrete biocultural significance.

Quickening

In many cultures, a baby's life is said to begin when its birth mother can feel it moving in her womb. This can begin to happen about four months into a pregnancy, by clinical calculations; it happens when the woman is entering the second trimester (biomedicine divides pregnancy into three three-month phases). At that point, a fetus begins to move in a way that can be felt by the person carrying it in her body. We call that **quickening**. The Catholic Church has called it 'ensoulment.' In Catholicism, people traditionally believed that a soul would not enter a fetus until after it had formed up enough to have a humanlike look to its organism—which generally occurs when the first trimester has ended.

However labeled and interpreted, the quickening is a universal experience, and many cultures interpret it as the point in time when the fetus becomes alive. From this point of view, an early-term abortion is not a cultural and moral problem because what is being aborted is not 'human.' In fact, it was not until the late 1800s that the Catholic Church became sure itself that abortion before the quickening could be culturally and morally problematic. At that time in European history, birthrates were declining among certain classes. Church leaders decided, after some wavering on the point, that ensoulment was simultaneous with conception. Abortion at any time during a pregnancy then became nearly impossible to justify. Spiritual truths aside, as seen within the political and economic context of the times, this decision by church leaders made sense.

Other Beginnings

Many cultures today believe that human life does not begin until the quickening. Others put its commencement at the point of birth. Still others put it a certain number of days after the baby has come into the world and proven itself viable. In fact, many cultures do not even give a baby a name until the dangerous phase just after birth has passed, if the baby is still alive. Some cultures at that point will name the baby, considering it now a human being—or making it one via the performative act of baby naming. Some wait until the teeth have begun to come in to call a baby 'human.'

Here, we might consider the interesting and very biocultural question of how babies turn into (or are turned into) real people. In some cultures, a vast amount of work must be done in support of this process—biological and cultural work. We saw in Chapter 7 that, in babyhood, this work might involve early weaning to promote independence. It also could entail favoring breastfeeding; in Anthroposophy (an Anglo-European spiritual subculture descended from Theosophy), it is thought that breast milk awakens the child's spirit; it contains 'will' forces necessary to jump-start proper child growth and development (Steiner 2007, 153–154).

As informative as exploring the details of healthy child development in this and other worldviews might be, we were really asking the question of when life begins. It is a question with great salience in the United States today. However, others consider the question ridiculous: for them, life never begins or ends.

Buddhist Perspective

An example of this is found in the context of traditional Japanese Buddhism which, as William LaFleur describes it (1994), considers life as a wheel (see Figure 13.4). The soul is not on a linear journey from some starting point that finishes with death and perhaps an afterlife. Rather, its journey is circular; it is continually reincarnated. It moves back and forth between the realm of the sacred and the secular, 'real' world.

Souls stay in the river-filled, liquidy realm of the spirits until the time comes for them to be born in human form. At that point, they make a journey from the other world into this one. According to traditional Japanese Buddhism, as the child grows and ages, he or she becomes more and more solid. Becoming human, then, is a process of solidification. Children eventually lose their otherworldly nature, but it is not really until someone is teenaged that his or her humanity has solidified to the point that the spirit realm is forgotten.

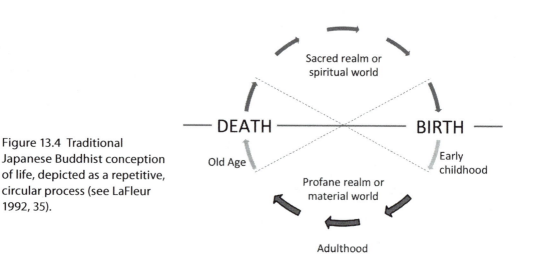

Figure 13.4 Traditional Japanese Buddhist conception of life, depicted as a repetitive, circular process (see LaFleur 1992, 35).

Adult life is mundane compared to the life of the child because adults are so solid. They have lost that quick connection to the other world. Nevertheless, later in life, we begin to lose some of our human solidity again. Over time, assuming that we do live a long life, we will begin the journey back into the other world, slowly rejoining it. Dementia thus can be taken merely as a sign of an inevitable progression. We come back in touch with the sacred world and eventually, upon human death, re-enter it completely, sooner or later to be born again. In this way, life is everlasting.

As LaFleur describes it, the sacred world is very liquidy. Rivers, streams, and springs are seen as conduits to that watery world. Often, if a woman miscarries, has an abortion, or has a stillborn child, that fetus or infant will be buried under the house. Even if there is not a real river under the house, the idea here is that water flowing below the floorboards, real or symbolic, will carry its soul back to the sacred realm safely. Moreover, there are certain rituals to ensure this, and to protect the soul of that unborn child so that when the time is right for it to actually enter the human world, it will come back, and it will have a good life.

That ritual is called the *mizuko kuyō* ritual. Mizuko means 'water child' and refers directly to the unborn (the fetus, the aborted, the miscarried, the stillborn), whose soul is sent back to the watery spirit world. Kuyō means 'to offer' or 'to nourish,' so the mizuko kuyō involves an offering meant to nourish or keep the mizuko. In other words, a mizuko kuyō is a memorial service for a water child. In it, an offering is made to the Mizuko *Jizō*, the guardian deity, sometimes spoken of as the womb of the earth, who protects not only pregnant women and water children but also travelers and pilgrims—anyone in the midst of making a passage.

Figure 13.5 Zōjō-ji (mini Jizō) in temple garden, Minato district, Tokyo, Japan. Used in memorial to the unborn. *CC-BY-SA-3.0.*

The shrine areas where these offerings are made are full of little doll-like statues (see Figure 13.5). Those represent the miscarried, stillborn, or aborted souls. Some have little red bibs or knit caps, just like a living infant would. Some have pinwheels in their hands as a living child might hold a toy. Placing one of these at a mizuko kuyō shrine and making associated offerings gives people a very concrete way to handle the grief that they feel upon losing a potential child and family member.

Mizuko kuyō does not exist just to make parents feel better. It is crucial for the protection of the water child; it helps protect it, and ensures as well that it will come back when the time is right. The thinking here is that bad timing underlies miscarriage and abortion. Within the family it simply is not the right time to bring another soul into the world. Maybe there is a very sick sibling who needs a lot of attention and help; the parents therefore would not have the time necessary to give a new baby the love and care that it needs. Maybe there is not enough money available, and so sending this soul back is a way of making sure that it can instead be born when the time is right and family finances are back in order.

In other words, abortion is not a final life-ending act but rather a simple delay. Abortion is, then, from the traditional Buddhist point of view, the right thing to do if a family or woman wishes to bring a child into this world properly. In this tradition and others where life is not linear—where it does not have a certain start point and a certain end point but moves in a circle—abortion promotes family values and serves as a very pro-social act (LaFleur 1994).

Family Building

Infertility

While abortion is, by definition, something done on purpose for various reasons, miscarriage (also sometimes called 'spontaneous' abortion) is not. Many times, when a conception miscarries—when the body ejects or rejects an unviable fetus—the woman whose body is involved does not even know. About half of all fertilized eggs miscarry. This frequently happens very quickly, before a woman even knows that she is or has been pregnant. Yet, about 10 to 20 percent of all known pregnancies also end in miscarriage. One or two out of ten is a notable amount.

Of course, providing figures like this, as if all populations are exactly the same, is misleading. Miscarriage happens more in populations that are undernourished or otherwise stressed. Miscarriage rates also tie in with infertility rates, which again are higher when health is taxed. When people's bodies are malnourished, for instance, they physically do not have the energy to support a pregnancy. In addition, they are less likely to be ovulating or producing viable sperm. Infertility therefore is linked with social structural issues. It ties in with poverty. It ties in with a lack of access to medical care as well, because certain diseases, if they go untreated, can lead to infertility. Today, although again this figure is generic, about 15 percent of all couples experience infertility.

Adoption

Because infertility is and has been so common, most cultures have certain strategies for how to deal with it. In many cases, these entail adoption or **fosterage** (fosterage is social but not legal adoption, sometimes temporary, yet often long-term). Adoption, legal or not, is in many cultures a viable, respectable, normal mode of family building. Remember that in many cultures people can adopt children who have other parents, too, and people who already have children also can do so: recall the Azawagh Arabs' milk kin customs.

Beyond adoption, people can contract or bring in a **surrogate** or stand-in mother or father to help build their families. There is a story about surrogacy in the Old Testament of the Bible, in the Book of Genesis; this story also features obliquely in the Quran. It is a story about a couple, Abraham and Sarah, who could not conceive a child. Sarah was thought to be barren (later she does bear a child, but that is another story). To solve the immediate dilemma—to generate for Abraham an heir (theirs was a patriarchal world)—Sarah's maidservant Hagar was called upon to have sexual intercourse with Abraham and bear a child for the couple. Hagar was Sarah's surrogate. She bore for Abraham Ishmael, who became the patriarch of the Ishmaelites, or Arabs. So surrogacy is important in all of the Abrahamic faiths.

Assistive Reproductive Technology

Today, another option exists in the form of ART, or assistive reproductive technology. One kind of ART is *in-vitro* fertilization (IVF), in which eggs from the genetic mother and sperm from the genetic father are physically extracted, mixed together in a Petri dish to promote fertilization, and any resulting embryos are implanted into a womb, either that of the genetic mother, that of the intended birth and adoptive mother, or that of a gestational surrogate. A gestational surrogate is simply a stand in; with gestational surrogacy as with old fashioned surrogacy, such as Hagar practiced, the baby is handed over to the family that requested or contracted for it.

The new technology is awesome. It also is very expensive. Moreover, the process is very time consuming and quite risky. Whosoever's body the eggs come from must ingest or inject hormones for a while to stimulate ovulation. Instead of the one or at most two eggs that normally ripen, if any do, in a typical menstrual cycle, a woman undergoing IVF might produce from five to twenty eggs. This in itself puts extra stress on the body and overstimulated, enlarged ovaries may result, leading to concurrent health problems. The operation to withdraw the eggs involves general anesthesia, which always entails a degree of risk. Emotionally, IVF also is a risky, stressful business, due to uncertain outcomes and the strength of the (culturally relative) desire for a genetically related baby that drives people to undertake the process.

Then, there are the risks to the person who will carry the fetus to term. For example, clinically assisted implantation entails a risk for infection and medical error that might lead to harm. Carrying the child to term puts certain stresses on the pregnant person's body, and there is no guarantee for the health of the expected child or the gestational mother, although surrogates generally are well monitored by clinicians and in some cases receive supplements to ensure a healthy diet and so forth. The scheduled cesarean section, often used with gestational surrogacy so that the receiving parent or parents can be present at the birth, has its own risk for complication list.

Despite these risks and others costs, there also are benefits to the IVF system. For example, a child is conceived and born, helping an infertile person or couple to fulfill a culturally underwritten dream of parenthood. Others benefit, too, for example via financial arrangements.

Culture's Role

By now readers know that culture is part of the family building equation. Normative cultural ideas about kinship affect our strategies for building families and help to generate our desires for particular kinds of families to begin with. They also affect legal regulation of family building strategies, including not only adoption but also regarding ARTs.

When a new ART comes on the scene, heated debate follows about who has the right to use it, and under what conditions. For instance, people ask if there should be age cut offs or behavioral standards imposed on applicants, whereby, for example, people who are over forty or who have engaged in prostitution or drug use or belittled the Queen in public might be denied access. Experts and lay people alike debate whether we should limit the use of ARTs to the normative familial context (and in mainstream America that would be a legally married male-female couple) or whether we should allow male-male or female-female or even multiply partnered spousal units and single people to use them. These decisions are made on a cultural basis, not a medical one.

Our normative cultural ideas about kinship and the family affect the choices we make in the face of infertility or when we want to have more children than we can without assistance. Today, the mainstream in America, more often than not, will opt to use ARTs when they can afford them because of cultural understandings about 'real' kinship as anchored in shared DNA.

These understandings are, in fact, a key driver in the expanding market for medical travel, in which people cross borders to seek out medical or dental treatments that they cannot access at home, whether because such treatments are too expensive or not yet done there or simply not legal in their home country. In terms of gestational surrogacy, India is one prime market for infertile Americans seeking babies (Kumar 2008; see also Sobo 2009).

From one perspective the emerging international gestational surrogacy market generates a win-win situation. Gestational surrogacy and the ART necessary to create it are very expensive in many countries. People who otherwise could not afford the process, say in the United States, can go elsewhere to get it for a fraction of their homeland cost. Using weighted average procedure prices, one research company put the average savings from the U.S. perspective for operations undergone overseas at about 85 percent (Keckley and Underwood 2008). Thus, medical travel makes parenthood an option for people who are not wealthy or who lack insurance coverage for the expense.

Moreover, international gestational surrogacy provides people overseas who otherwise may be struggling with a source of income. The women in India who hire their wombs out can use the money to put their children through school, provide food for the family, or even buy a house.

As usual, however, if a situation looks too good to be true, it probably is too good to be true. In this case, there are lots of risks. In addition to the health risks to the surrogate already mentioned, such as with cesarean operations, there may be increased risk of multiple pregnancies and all that they entail, because in order to get one embryo to stick, clinicians may be more likely to gamble in a surrogate's womb and deposit too many, as a form of insurance that at least one will survive.

Plus there may be social risks to the surrogate. In many cultures, this kind of service—or being in a situation in which one is forced to provide it to survive—is very much frowned upon. The fact that a woman is so ill-provided for that she must rent out her womb says something to others about her social network's (or her own) failings. It is a self-stigmatizing act. Psychic risks include this plus also the cost of coming to see and experience one's body as partible, or more as if a machine with interchangeable pieces than a holistically integrated human organism.

Population-wide, the practice has ramifications for demographic patterns. As ART access rises, the demand for children to adopt goes down. An increase may be noted in orphan or abandoned child populations.

Yet, for mainstream Americans, use of gestational surrogacy or other ARTs is often the first choice that comes to mind when seeking to build a family and simple sexual intercourse does not suffice. Despite the risks, it is the choice that makes cultural sense. Our mythic belief in the ability of DNA to create kinship is so powerful that it can drive people toward ARTs and away from adoption on their family building journey.

In non-Western contexts or where DNA is not the main substance of kinship, adoption is much easier—and it really fits well for family building in more open, flexible kinship systems—the kind that allow people to construct kinship additively postnatally from substances such as breast milk, food, nurturance, and so forth. In that case, a resort to ART makes no sense at all. People will not take those risks if they can just adopt and create consubstantiality at home, cheaply, safely, and easily. They will only do it if raised in a culture that valorizes the DNA connection. In contrast, the Hua would say that those who favor ARTs over adoption are out of their minds.

It is true that we do have a tradition of adoption in the West: kinship can be created affinally. Nevertheless, culturally speaking, and as historical records demonstrate, this kind of kinship has been seen as second rate. In past generations, adoptions often were kept secret because of the potential stigma and for fear that the parent-child relation would somehow be compromised if the adopted child knew—or that his or her relationship with any siblings might be damaged. Adoptive children and parents were put on the defensive by the normative standard that lineal kin be genetically related (the same was true in earlier historical periods when bloodline standards reigned).

Although secretiveness regarding adoptions does seem to be on the wane, we still see culture-induced defensiveness in the fact that adoption is seen as something to be remarked upon at all. We see it when adoptive parents and offspring justify their relationships as if 'real,' saying things like, "My mom and dad are like real parents to me. I think of them as my real parents. They are my real parents." The fact that such statements are even made is culturally revealing.

Diverse Modes of Meeting the Drive to Connect

American defenses of the relatedness of adoptees and adoptors reveal something else also: even while we overtly prioritize genetic kin ties, we hold additional, contrastive understandings regarding kinship that have much in common with other cultures—cultures that prioritize diverse sources for kinship, including nurturance. In this they bear witness to one of the key traits of humans: our need for, and our ability and impulse to build, social connections. It is the case that there are numerous ways to build and explain these connections; possible sources for consubstantiality, for instance, are many when the entire ethnographic record is taken into account. Ideas about when connections can be or are built, or how long they might endure, also vary. The biocultural diversity entailed in and by these differences notwithstanding, all of humanity shares a universal and inherent drive for communion.

Conclusion
Respecting Connections

In Part I of the book, our focus was on how we adapt to the geographic or natural environment, with its predators, prey, and so forth. We paid special attention to co-evolution, the related benefits of biological diversity, and the role of niche construction. We saw that many adaptations fed into each other in our evolution, synergistically shaping our species. In Part II, we began to think about biocultural diversity spurred by the many changes brought about by our ancestors after they spread across the globe—changes to both the geographic environment and the social one—particularly once we settled down. These changes came about because of how we decided to manipulate the environment in our quest for more food, and how we decided to organize ourselves when we intensified our subsistence strategies. These changes had benefits, but they had certain problems, too. In the last part of the book, our focus shifted to a consideration of how we adapt to culture itself—to the beliefs, values, and ideals we hold, and which in many ways are imprinted upon and in the body. In other words, this book has focused on human adaptation to geographically, socially, and culturally generated selective ('environmental') pressures and the biocultural diversities that result.

After establishing the value of a holistic, systems approach, the book first tackled its explanatory task from a biological perspective, taking questions related to genetic inheritance as a portal of entry into a sphere of inquiry that is simultaneously biological—physiological, anatomical, genetic, and so forth—and at the same time cultural—social, political, economic, historical, ideological, and so on. We started on the biological side because some of the key concepts used today in the social sciences were developed first in that arena (for example, the theory of natural selection). By the last chapter, however, we had in effect come full circle: we ended by investigating inheritance again, but this time in terms of the cultural ideas and practices of kinship and related issues.

We might instead have begun the exposition at this cultural end. In European anthropology, the comparative study of kinship came first; it fostered anthropology's emergence as a discipline in its own right. In the United States anthropology emerged in response to racism, for instance via Franz Boas, and had a very physical, biological focus initially, but in Europe, origin stories hark back to William Halse R. Rivers and his genealogical method, used perhaps most famously on the 1898 Torres Straits Expedition. I bring this up to reinforce the argument that multifaceted issues can and should be approached from multiple entryways. The placement of the kinship chapter at the end, and therefore near to Chapter 1 in the circle the chapters create, represents a final request for holistic thinking.

Having come to this point—having been through all three sections—four key propositions should be obvious to all:

- Biology and culture co-create each other and the human experience
- Holism and systems thinking are useful
- Variation is good
- We are all connected

Before closing, then, I shall review these propositions one by one. In doing so, I shall also remind readers of some areas that need attention, both scientifically and humanistically.

Biology and Culture Co-create Each Other and the Human Experience

A key question asked throughout the text is: How do culture and biology interact? *Biology and culture interact synergistically, creating together something that each cannot create on its own.* Biology and culture (once culture emerged, that is) reinforce one another; they co-create one another; they produce what we know of as the human experience.

While some of the things we have studied, such as tattooing or the distribution of skin pigmentation between populations, imply that biocultural interaction occurs only where we can see it, much of the interaction between biology and culture takes place inside of the body. A lot of the important effects of the biology-culture interaction happen out of sight, at the cellular and even biochemical level deep inside human beings. Think here, for instance, about some of what we learned when studying the allostatic load and epigenetics. We should always stay aware of the invisible, which sometimes is more important than the seen. We should also stay aware of the importance of regional difference; what we learned about geographic clines highlighted the error of lumping groups of people together too broadly. Significant differences often occur on a much smaller scale (as well as across different lines) than those who are wedded to the old-time conception of race would believe.

Holism and Systems Thinking Are Useful

The book's focus on the synergy between biology and culture directly reflects anthropology's celebrated holistic perspective. It also demonstrates anthropology's systems perspective, which proposes that unless one understands how a system works—and in this case our interest is human systems—one really cannot understand in any intelligent fashion what's going on in that system. It makes no sense, from anthropology's holistic, systems-oriented viewpoint, to examine the parts of our existence on their own, or stilled from action. Rather, we should examine the parts working together, dynamically, in movement. We should be interested in the relationships between parts of human systems, and in what emerges out of the interactions between those parts. Whatever outcome may be achieved is achieved by the system as a unit or as a whole. Applying this insight to investigations of such varied phenomena as educational achievement gaps or differences in disease rates or athletic and other such capacities between defined groups of people can fuel discoveries regarding such diversity.

Not only are human societies systems, they're complex ones; we used the concept of the complex adaptive system to gain a toehold in our understanding of how human systems work. Understanding how systems work is key to constructively changing them. In a complex adaptive system, the parts of the system can adapt—they can learn, altering the system's direction. A system that can adapt is better able to survive than a mechanical (non-adaptive) system when its environment changes.

Our whole investigation of the dynamics of biocultural diversity was informed by an interest in adaptation. For much of the book, adaptation was measured with the index variable of health. The health of a population is a good indicator of how well or how poorly that population is adapted to the context in which it lives.

With that in mind, the book's key lesson is basically this: neither biology nor culture can stand alone. *Culture and biology are related in a dynamic, synergistic way, co-creating one another.* Our biology would not be what it is today had culture not influenced it, and our culture would not be what it is if biology had not had its influences as well. Despite the implications of the fact that English only gives me separate words here (biology, culture), they and their influences are conjoined.

Long ago, that fact was obvious. But over the last few hundred years, as universities grew and multiplied, disciplinary walls were created. Scholars carved off and laid claim to different aspects of existence, building ring fences around them. As Rudolf Steiner explains, the common approach in science has been to pull the fabric of life apart in search of abstractions. The legacy of conventionally bounded disciplines reflects this on a general level. However, as Steiner further notes—and here the emphasis is his—"*To explain a thing, to make it intelligible*, means nothing else than to place it into the context from which it has been torn.... A thing cut off from the world-whole does not exist" (2008 [1894], 74).

The fallacy of reductionism aside, habitually viewing things—for instance, facets of human diversity—as artificially bounded, abstractable phenomena existing in isolation can leave us less able to draw back and take a wide-angled view of things. It can blind us to the connections that do exist, for instance supporting materialist, capitalist logic that says anything goes when profit is at stake. A more holistic way of thinking—a unified point of view—may help us to realize a more humanitarian future in the long run as well as promoting, in the nearer term, a more realistic and comprehensive understanding of human diversity.

Variation Is Good

We know much more today than we used to about the range of variations between human populations. For instance, as examples included in this book demonstrate, we understand now how many variations occur—and we know now the value of holism for increasing the richness of this understanding. But there is still much we do not know, including in regard to the full range of human plasticity, and the enablers of and barriers to (as well as the limits on) various forms of this plasticity. Another area among many ripe for investigation is the generation-to-generation transfer of epigenetically acquired traits. All this makes biocultural studies a fertile field for future research.

One axiom of such research is that variation is good. For example, genetic variation helps to ensure that an ample number of able individuals reproduce, helping the group as a whole to survive into the next generation. Social and cultural variation, too, are helpful in terms of our population's survival; they provide us with a wide range of ways to be—some of which of course are more adaptive (or, better fitted to our environment) than others. Tamping out variation, through whatever means, would be to put all our eggs in one basket. As experienced investors and innovators know, this is not a good idea. It lessens our resilience and creativity in many ways. Seen in this light, biocultural diversity's helpfulness comes clear. We've learned throughout the book about diverse ways of organizing social groups and of cultural practices that do not make sense to outsiders but which are perfectly logical when viewed from the inside perspective. We have put aside ethnocentrism in favor of such cultural relativism so that we, too, might understand— and benefit from viewing—foreign practices in their own light.

Of course, just because something 'makes sense' in cultural context does not make it right. Some practices may actually be better than others in terms of allowing all people to optimize their developmental potential. One force holding people back from this in some areas is the global political economy, which affords some people more benefits than others— benefits that come at a cost to those who do not have them. Oftentimes, the costs are embedded in people's bodies, as through job-related injuries or diseases stemming from the chronic stress brought on by the conditions of low-wage labor, unsafe housing, lack of educational and health care access, and the like.

We Are All Connected

Lest I end this book on a melancholy note, I must restate that the global system is global; that is, we are part of it. The lives of the global poor and even of the poor in one's own nation, not to mention one's neighborhood, often seem as if they are lived in completely different worlds from the lives of the middle class, let alone the rich. These worlds are not separate, however; they are intimately connected. They are part of one system and exist in dynamic relation to one another. An action in one part of the system can have effects—even large ones—in another of its parts.

As a final reminder of our participation in complex adaptive systems and an example of how small changes can lead to big ones, let us return to consider the food movements gaining traction today (see Chapter 7). By fostering, on small farms, the kinds of multi-stranded relationships that were prevalent prior to industrial agriculture's onset between biotic populations and the earth, air, water, and sun, and by striving to bridge the contemporary divide between consumers and producers, these movements can help offset some of the ills we humans have created over recent centuries as a downside of factory logic.

It is the case that organic food produced by individual or small-holding polycultural farmers can be hard to find and expensive to buy (finding plain 'organic' food is much less a problem these days due to its industrialization). However research suggests that, as this subsistence style catches

Figure 14.1
Cecilia Joaquin gathering seeds with traditional Coast Pomo seed beater and burden basket. Photograph by Edward S. Curtis, c. 1924. *Library of Congress Prints and Photographs Division LC-USZ62-116525, cph 3c16525.*

on, the benefits to society will include fewer toxic pollutants, less use of fossil fuels, and increased floral and faunal biodiversity. Health expenditures may be lowered, too, and health outcomes enhanced.

Potential health gains may extend beyond those brought about by the better nutrient profiles that many non-industrial crops offer. For instance, without the irrigation ditches that large-scale monocropping can demand, the risk for malaria (in regions that have it) may go down and with that, over the long term, so, too, may the frequency of the sickle cell allele. Further, if consumers become more involved in food production, even if only through contributing a few hours of labor each week as part of a cooperative agreement, risks related to physical inactivity could be lowered. Gains in people's sense of connectedness and overall well-being also can occur, with concurrent positive effects on populations. Right now gains are local, but members of most branches of the new food movements—biodynamic farmers, for instance—demand a social justice approach that targets the globe (Biodynamic Farming and Gardening Association 2012).

A good case in point involves *Navdanya*, a movement for biodiversity conservation and farmers' rights based in India. The name, Navdanya, refers to nine seeds that are part of India's agricultural legacy. Seeds are life sources; properly fostered they end up as whole plants, which naturally produce their own seeds for self-regeneration (see Figure 14.1).

This is not so, of course, for industrial hybrids, which is one problem that Navdanya was established to stand against: after large agribusiness interests gained prominence, farmers once able to supply their own seeds (from last year's produce) now had to buy them. They also had to buy expensive pesticides, to safeguard the non-indigenous crops. The hazards entailed went well beyond exposing the soil, water, flora, fauna, farmers, and, by extension, farmers' families to toxins. They included indebting farmers, too, and inducing stress-, malnutrition-, and disease burden-related impairments in farmers' children—impairments that can reduce cognitive capacities and so reinforce a cycle of poverty (see Chapters 9 and 10). The pesticides also have been linked to a titanic wave of farmer suicides, which itself has had numerous down-the-line effects on households and communities. Navdanya would turn farmers who have grown dependent on and been impoverished by commercially purchased seeds and chemicals back toward self-sustainable organic methods.

Having said that, when Navdanya was conceived, previously long-standing farming techniques as well as local crops had come under threat of extinction, having been cast aside as farmers turned toward modern methods. To promote food sovereignty, Navdanya had to work to keep old traditions alive, for example by training seed keepers and establishing community seed banks where heritage seed varieties could be cultivated, stored, and distributed. Within twenty years, there were fifty-five such seed banks (seed banks have enjoyed rising prominence in other nations too; see Figure 14.2). India has also passed some legislation protecting small farmers. Navdanya and its instigator, ecologist Vandana Shiva, have gained the world's attention. They serve as models for others who seek similar ends (Navdanya 2009).

Figure 14.2 Sculpture by Peter Randall-Page, "Inner Compulsion"; commemorates the seed collection of the Millennium Seed Bank (*in background*), Wakehurst Place, West Sussex. *Photo by Jim Linwood, 2010, CC-BY-2.0.*

The links between human biocultural diversity and small-scale organic farming take sophistication to see. However, readers of this book by now can recognize that subsistence arrangements and the larger political economies that they are part of underwrite many of the differences seen between populations today. Incorporating into our subsistence tool-kit polyculture such as Navdanya or the biodynamic approach recommends could offset negative, health-sapping differences as well as promote a more humanistic social order. Because this kind of farming requires regional sensitivity, so that indigenous or at least geographically appropriate crops and livestock are emphasized, helpful biocultural differences also can, in theory, be maintained.

Some may assume that the kinds of broad-based changes necessary to support the widespread adoption of models like Navdanya's can never happen because the current way of life is too well-entrenched. However, more extraordinary changes have been witnessed, such as those entailed in paradigm shifts in science; an example we studied concerned epigenetics (see Chapter 3). Comparably, shifts in how people understood biocultural differences related to geographic ancestry ('race') supported the passage of the Thirteenth Amendment (outlawing slavery), the military's integration after World War II, school desegregation in 1954, and the civil rights protections enacted in the mid-1960s. These shifts required remodeling large-scale and heavily institutionalized political-economic systems of inequality. Changing such systems seemed as far-fetched to some people back then as the idea of putting a person on the moon. Yet change they did, and as relationships between and within populations shifted, so, too, did the landscape of biocultural adaptation and diversity.

As history has shown repeatedly, and as we know from studying some of the mechanisms of human biocultural adaptation, change does happen. Given the right conditions, even the smallest of variations can turn out to have huge ramifications over the course of time. It behooves us to pay attention to this—and to do as much anticipatory work as is possible to help ensure that whatever emerges is to our world's benefit. Even changes such as those discussed above in regard to food could have negative consequences if all angles are not proactively considered and dealt with.

My point here is that our future actions, and insights gained through the application of the unified approach promoted in this book, can have just as much, if not more, impact as the actions and discoveries of those who went before us. We can strive to set right mistakes made in the past, and we can work proactively toward the future we desire. We can scientifically investigate the puzzles of human difference from a more holistic perspective that, because of its comprehensive emphasis, enhances and adds to the body of knowledge that precedes us.

A unified approach to the magnificent array of human variation reveals that our experience as human beings comes out of the biology-culture relationship. Nature and nurture together create us as we are. Diverse as human populations may be, we all live in the world together. Let us leverage what we have learned about biocultural diversity's causes and effects in a concerted effort to optimize humankind's potential and thereby the potential of the larger systems we are part of.

Dynamics of Human Biocultural Diversity: A Unified Approach by Elisa J. Sobo, 305–313.

Appendix

Selected Websites for Supplemental Multimedia Learning Resources

This short list contains a small sample of URLs housing material that might be used in tandem with this book for accelerating exploration and understanding. Specific citations for complementary material also are found in References.

American Anthropological Association (AAA)	www.aaanet.org
British Broadcasting Corporation (BBC)	news.bbc.co.uk
Economist—Science and Technology	www.economist.com/science-technology
Evolution (PBS)	www.pbs.org/wgbh/evolution
New York Times	www.nytimes.com
New York Times Video	video.on.nytimes.com
Nova—Evolution (PBS)	www.pbs.org/wgbh/nova/beta/evolution
Nova—Science Now (PBS)	www.pbs.org/wgbh/nova/sciencenow
Palomar College Anthropology Tutorials	anthro.palomar.edu/tutorials/cultural.htm
Race—The Power of an Illusion (PBS)	www.pbs.org/race/000_General/000_00-Home.htm
Race: Are We So Different? (AAA)	www.understandingrace.org/home.html
Smithsonian— Human Origins	humanorigins.si.edu
TEDTalks	www.ted.com
Understanding Evolution— UC Berkeley	evolution.berkeley.edu
Unnatural Causes (California Newsreel)	www.unnaturalcauses.org
Webquest.org	webquest.org/index.php

Glossary

A

Abiotic—non-living chemical or physical factors in an environment, such as soil, atmospheric gasses, sunshine, and water

Acclimatization or *acclimation*—short-term, reversible changes in an organism's biology, such as when a summertime beachgoer develops a seasonal tan

Acute disease—a sudden onset ailment; comes and goes (or kills) quickly

Adaptation—a responsive, survival-enhancing change in a system brought about in reaction to some kind of atypical or new and stress-producing change in the environment

Adaptive immunity—immunity brought on through vaccination or inoculation

Affiliative—an act or mark (e.g., a body decoration or modification) that symbolizes one's affiliation or identification with a particular culture or subculture; antonym is 'disaffiliative'

Affinal kin—kin related by law, such as through adoption or marriage in the United States

Agency—the ability to impose one's will or make significant choices

Agent—pathogen, substance, or process that causes morbidity and/or mortality

Agriculture—settled farming, practiced with plows, irrigation, and/or fertilizers

Alleles—variants of a single gene; see also 'polymorphic'

Allocare—the provision of childcare services to children who are not one's biological offspring; literally, 'other care'

Allostasis—the process of re-creating homeostasis by changing the body's initial set points or 'factory settings' to accommodate chronic stress

Allostatic load—cumulative multi-system physiological dysregulation resulting from chronic stress; a higher allostatic load places an organism at a higher risk for poor health

Androgynous—expressing both masculine and feminine characteristics

Anomaly—something that does not fit into preconceived cultural categories, is unclassifiable, or combines traits from two or more categorical types (the latter also is sometimes termed a 'monster'); an anomaly has power because it disturbs, or represents a disturbance in, our culturally influenced sense of order

Anthropogenic—humanly created or generated; often applied to describe diseases or problems related to human-made changes to the physical environment, such as via agricultural practices

Anthropology—study of humankind; holistic in perspective

Antibodies—special proteins that can lock, selectively, onto the invading cells' antigens and effectively disarm them

Antigens—biochemical substances that mark cells as belonging to ourselves or not

Archaeological record—all material culture or artifacts and other remains of historic and pre-historic societies

Artificial selection—purposeful breeding of plants or animals, by humans, with the intention of producing in their offspring particular traits

Asexual reproduction—self-replication, for instance by splitting, so that each generation is a duplicate of the last; reproduction involving no sex; see also 'cloning'

Assimilation hypothesis—holds that populations of *Homo sapiens* who left Africa in earlier waves (see 'out of Africa model') were absorbed or assimilated into newer populations

Attenuation—when a pathogen evolves, via natural selection, to have less of an impact on its host, and its host, as a result, can carry or spread it for longer

Avunculate—matrilineal relationship between a man and his sister's children

B

Balanced polymorphism—exists when selective pressure for one form of an allele is balanced or offset by selective pressure against that form, so that both alleles persist in a population, having achieved some kind of balance

Band—egalitarian, family-based socio-political group; leadership is achievement based

Bioavailability—quality of a food item affecting the degree to which an eater can extract and make use of nutrients; bioavailability is affected by such things as food storage conditions, cooking processes, and juxtapositions with other food items

Biocultural diversity—all population-based human variation generated in or reflecting the dynamic, synergistic communion of biology and culture

Biological determinism—the belief that biology alone determines one's capacities and characteristics (often used in opposition to 'cultural determinism')

Biotic—alive; a living component of an ecosystem, such as a plant or animal species

C

Chiefdom—hierarchical socio-political group run by a hereditary bureaucracy, in which people inherit their positions

Cholera—potentially fatal, frequently epidemic acute and often water-borne bacterial infection of the small intestine; symptoms include copious amounts of watery (i.e., often fatally dehydrating) diarrhea

Chromosomes—organized, structured packages of DNA found in the center or nucleus of each cell and passed along to offspring during reproduction

Chronic disease—lasts over a long time span

Classificatory kin—fictive kin; people who are referred to with kin terms and said to be related but known by everyone not to be

Cline—see 'geographic cline'

Cloning—self-replication, for instance by splitting, so that each generation is a duplicate of the last; reproduction involving no sex; see also 'asexual reproduction'

Closed societies—societies in which a person's social status is ascribed; often it is based on birth

Colostrum—antibody-packed liquid that comes out of breasts prior to breast milk in a new mother

Commensality—eating together to create and cement a bond; sometimes this bond is literally one of shared substance

Communitas—sense of bondedness or community loyalty that people who go through a rite of passage together have toward one another

Comparative method—entails viewing each culture in comparison with others or from a comparative perspective so that universals can be derived

Complex adaptive system—system or network of dynamically interrelated parts between which information can flow; this system can change itself adaptively in response to changes in its environment (compare to 'mechanical system')

Complex society—chiefdom or state; heterogeneous in structure (refers to social organization, not culture)

Concordance—when features or traits co-occur with predictable regularity

Consanguineal kin—kin who share blood or are from the same bloodline; no other substance but blood is implicated in consanguineality, which is therefore a subset of consubstantiality (see 'consubstantial kin')

Consubstantial kin—kin related through shared substance, whether immaterial (e.g., nurturance, love) or material (e.g., bodily fluids such as breast milk; DNA, blood, but see 'consanguineal kin')

Co-sleeping—purposefully sleeping in close proximity, for instance in the same room (but not necessarily the same bed) as per cultural recommendations

Couvade—male participation in pregnancy, sometimes as demonstrated through the male experience of food cravings or morning sickness

Cultural consonance—the degree to which one's lifestyle fits with the lifestyle that one's culture recommends and that one thereby aspires to

Cultural determinism—the belief that culture alone determines one's capacities and characteristics (often used in opposition to 'biological determinism')

Cultural Revolution—dramatic shift in the archeological record occurring about 75,000 years ago indicating the onset of true behavioral modernity in *Homo sapiens sapiens*, including full-fledged capacity for culturally supported cooperative social life; see also 'Upper Paleolithic Revolution'

Culture—a uniquely human adaptive strategy; the totality of each human group's shared, learned heritage; this includes social, political, economic, religious, cosmological, linguistic, health promotion, and other systems

Culture-bound syndromes—culturally unique, culturally named conditions that bring together, as syndromes, a variety of symptoms; often linked to stress.

Cystic fibrosis transmembrane conductance regulators (CFTR)—see 'transmembrane regulators'

Cystic fibrosis—recessive genetic disease that disrupts the body's ability to absorb nutrients, breathe, and sweat; generally fatal in childhood if untreated

D

Developmental adjustment—a responsive change in growth or biochemical processes during development that occurs in reaction to environmental conditions

Developmental—related to or occurs during the time that an individual organism is growing or maturing

Diabetes—non-infectious chronic disease in which the body's ability to regulate glucose or blood sugar via insulin is compromised; can be controlled through diet, exercise, and medication; found more frequently in populations with higher levels of obesity and a more sedentary lifestyle

Dietary staple—food (usually plant-based) that a group eats a lot of; examples for various cultures are bread, tortillas, mongongo nuts, and rice

Disease ecology—theoretical framework that focuses on the germ, the immediate or proximal environment in which it lives, and the context in which germs are spread

Division of labor—extra-familial economic specialization, in which different groups within a society do different jobs and are therefore interdependent

DNA (Deoxyribonucleic acid)—biological molecule containing genetic information; each strand of DNA is essentially a string a genes

Dominant alleles—make their own instructions take precedence over recessive alleles so that only they are expressed as a trait; a dominant trait is a trait expressed in this way

E

Ecological selection—an increase in a given trait's frequency in the next generation's gene pool that happens when a change in the environment makes that trait advantageous; natural selection's workhorse

Ecosystems—balanced systems comprising multiple abiotic or non-living materials and biotic or living populations, each occupying a particular niche, coexisting in a balanced way, via energy flows, so that species' population numbers and volumes of abiotic materials hold steady even as time passes

Embodiment—the literal 'making physical' of culture

Emergence—a novel property or pattern that comes about as the result of interactions between parts of a system that aim to keep the system working and, in doing so, lead to something entirely new and otherwise unpredictable

Enamel hypoplasia—a dental condition; a sign of nutritional stress; there is less enamel in some places on the teeth than there would be on healthy, evenly-nourished teeth

Endemic—disease that is local to or well-established in a particular region and exists in balance with the population that hosts it

Environmental pressure—anything in the environment that reduces a population's ability to function or puts a damper on its potential; includes disease, deprivation (through drought, famine, and so forth), disaster (flooding, earthquake), and an increase in predators

Environmental reservoir—abiotic substance that can hold agents so that hosts come in contact with them (e.g., water, soil)

Epidemic—disease that spreads over a population with great speed

Epidemiological polarization—when the epidemiological profile of the poor entails high levels of mortality from infectious disease and lots of death at young ages, while the rich live longer and die later from non-infectious diseases

Epidemiological profile—a profile or picture of a given group describing what diseases and other health challenges they are experiencing or have experienced, and to what degree

Epidemiological transition—shift from an epidemiological profile highlighting infectious diseases with high mortality rates to one in which non-infectious or chronic diseases with low mortality but high morbidity rates feature centrally

Epidemiological triangle—triangle representing the relationship between the host, environment, and agent

Epidemiology—study of disease distribution and its determinants

Epigenetic events—biochemical processes in the epigenome that turn genes on or off or otherwise significantly affect their expression

Epigenetics—the study of interactions between genes and their environment that bring the phenotype into being (as per Waddington 1942)

Epigenome—a layer of biochemical processes overlying the genes and affecting their expression

Ethnicity—facet of identity related to shared national or regional origins and shared culture

Ethnocentrism—putting one's own culture at the center of any interpretation; viewing other cultures through one's own culture's lens

Ethnographic record—all ethnographic accounts, old and new, taken together

Ethnography—written account of one particular culture, traditionally in book format

Ethnology—comparative study of cultures

Ethos—a culture's overall worldview or fundamental values

Evolution—change, in any direction for any reason

Exposure—being in close proximity to a chemical, pathogen (germ), radioactivity, or extremes of weather.

F

Fauna—animals and other non-plant life forms (often used in the phrase 'flora and fauna')

Fight or flight reaction—when a stressed individual is immediately prepared via internal changes (a stress response) for a fight or for fleeing the scene; thought to aid species survival; see also 'stress'

Fitness—a measure of adaptation related to mortality and fertility; when comparing two groups, lower mortality with higher fertility is an indicator that one group is better fitted or adapted to an environment than the other

Flora—plants (often used in the phrase 'flora and fauna')

Foraging—gathering vegetable and animal or other foods directly, also known as hunter-gatherer lifestyle; often, foragers are nomadic, at least seasonally; about two-thirds of the forager diet is plant-based

Fosterage—social but not legal adoption, sometimes temporary but often long-term

G

Gametes—reproductive cells (eggs, sperm)

Gender—culturally recommended scripts for enacting masculinity or femininity; cultural ideas about what tangible sex differences mean

Gene—a discrete sequence of Deoxyribonucleic acid (DNA) that contains (stores or codes) the recipe for a particular protein, or regulates the expression of protein-coding genes; genes are passed along from parent to offspring through the process of reproduction

Gene flow—genetic evolution due to migration, when one group's genes flow into another's gene pool

Gene pool—the entire population's sum total of genes

Genetic adaptation—an adaptive change in the frequency of a given gene or genes in the gene pool from generation to generation caused by natural selection; a specialized form of genetic evolution

Genetic drift—genetic evolution due to random chance

Genetic evolution—a change in the frequency of a given gene or genes in the gene pool from generation to generation caused by a variety of mechanisms, including mutation, genetic drift, gene flow, and natural selection (genetic adaptation is a particular form of genetic evolution)

Genome—the entire stock of genetic information carried by a given species

Genotype—the set of genes an individual carries (as opposed to their manifestation)

Genus—a broad subfamily of organisms; each genus includes or is made up of various species

Geographic cline—incrementally or gradually changing distributions of traits over geographic regions, which are related to incrementally changing environmental pressures or challenges

H

Hemoglobin—oxygen-binding or carrying molecules found in red blood cells

Heterozygous—refers to inherited gene pairs; occurs when a person has two different alleles for a pair

Historical particularism—a viewpoint in which, because different societies have different histories, their cultures are therefore unique and particular (compare to 'unilineal evolution')

HIV/AIDS (Human immunodeficiency virus/Acquired immune deficiency syndrome)—AIDS is a chronic, potentially fatal viral infection of the immune system; HIV, the virus that causes it, is spread via direct contact with bodily fluids; death generally comes through opportunistic infections such as by pneumonia or tuberculosis

Holism—a perspective that views parts of a system within the context provided by the system as a whole; holds that single parts of a system cannot be understood in isolation and that the whole is more than the sum of its parts

Holon—a system that is simultaneously a whole unto itself and a part of something larger

Homeostasis—literally, steady state; a balance achieved when small changes are made that do not notably alter the system but instead allow it to run as it has been running

Homo erectus/ergaster—literally, 'erect man'; ancestral human-like species that stood erect and walked habitually on two feet; emerged about 1.5 million years ago

Homo habilis—literally, 'handy man'; ancestral human-like species that systematically made and used tools; emerged about 2.5 million years ago

Homo sapiens—literally, 'wise' or 'knowing man'; fully anatomically modern human beings; emerged 150,000 years ago

Homo sapiens sapiens—behaviorally and anatomically modern human sub-species to which contemporary human beings belong; signs of the full emergence of this sub-species as culture-carrying beings are seen in the archaeological record dated to about 75,000 years ago

Homozygous—refers to inherited gene pairs; occurs when a person has two identical alleles for a pair

Horticulture—gardening lifestyle (no irrigation, no plows, no fertilizer, no permanent settlements)

Host—individual organism sickened by an agent's active presence

Humanity—behaviorally as well as anatomically modern humans; see also '*Homo sapiens sapiens*'

Hunter-gatherer—see 'foraging'

Hygiene hypothesis—holds that the immune systems of children exposed to more microbes have greater tolerance for the irritants that trigger asthma and allergies

I

Index trait—a trait that indexes a potential mating partner's fitness relative to others in the group; in sexual selection, organisms rely on index traits (albeit generally not consciously) to help insure the viability of the next generation

Industrial melanism—when industrial processes darken the environment, leading to increased fitness for members of a species in which more melanic pigment is expressed and so supporting the natural selection of the melanic trait

Industrialized agriculture—farming (often of one crop only) for profit, rather than food, using highly mechanized means

Infectious disease—a disease resulting from the presence and activity of a pathogenic microbial agent that can be communicated from one person to another (contrast with non-infectious diseases, which involve no agents or germs)

Innate immunity—genetically inherited immunity

Intensification—the process of doing more to get food; more intensified subsistence strategies (e.g., agriculture) manipulate or interfere with the environment more than less intensive strategies (e.g., scavenging) do

L

Lactose tolerance—the ability to digest milk sugars (lactose) until late in life

Lamarckism—the idea that acquired characteristics can be inherited (named after naturalist Jean-Baptiste Lamarck)

Leukocytes—white blood cells; key to good immune system functioning

Liminal—occupying a threshold; standing in limbo between social statuses; the middle phase of a rite of passage

Lineage—descent group with shared ancestry

Lymphocytes—those leukocytes or white blood cells that help our bodies remember and recognize previous invaders and support our bodies in destroying them

M

Macroevolution—genetic adaptation that brings about whole new species

Macronutrients—the three key nutrients: proteins, carbohydrates, and fats

Malaria—a disease caused proximally by parasites belonging to the genus *Plasmodium* and carried by the *Anopholes* mosquito; entails cyclical high fevers, headaches, and often death in humans; one of the top ten causes of death worldwide

Market exchange—impersonal exchanges, often involving money; goods have no links to their producers; once a good is paid for no further obligation ensues

Matrilineage—descent group in which heredity is figured through the female lineage.

Meaning Response—self healing attributable to knowledge or belief (to the cultural meaning) that a practice or process carries; see also 'placebo effect'

Mechanical system—system of dynamically interrelated parts that cannot change itself or adapt to changes in its environment (compare to 'complex adaptive system')

Metabolism—the biochemical process of breaking down and repurposing food components for bodily use

Microevoluton—genetic adaptation that is not major enough to lead to speciation (the emergence of a new species)

Micronutrients—vitamins and minerals without which the biochemical processes entailed in fueling human life cannot happen

Monogenic—traits traceable to just one gene; most traits, in contrast, are 'polygenic'

Morbidity—sickness

Mortality—death

Multiple origins model—holds that *Homo sapiens sapiens* evolved concurrently, in a convergent fashion, in many regions and perhaps (but not necessarily) with some gene flow between regions (also sometimes called 'regional continuity' or 'multiregional' model; contrasts with the 'single origin' or 'out of Africa' model)

Mutation—genetic miscopying, often due to some kind of exposure, for instance to a toxic chemical, or radiation, or a particular virus

N

Natural selection—the key mechanism of genetic evolution; a process by which the genes for expressed traits that happen to give an organism an adaptive advantage for survival under given environmental conditions are therefore more likely to be found in the next generation's gene pool, if the same environmental conditions persist; descent with modification

Niche—species-specific way of making a living or subsisting

Niche construction—when organisms, through biological and behavioral processes, act upon (construct or modify) the environmental niches that they (and other organisms) occupy just as those niches act upon them

Nonconcordance—when things (features, traits) do not co-occur with predictable regularity; often referenced to dismiss claims that biological race exists

Nonlinear—when ramifications of an event do not follow from the event in a predictable manner but instead occur as surprises that lead to surprises of their own

Non-random mating—selective mating, driven by conscious rules or unconscious impulses; happens in human populations for instance when kinship or class-related rules or ethnic boundaries limit who can build a family with whom; see also 'sexual selection'

Normal science—scientific practice that is driven by a shared, paradigmatic theory about the world that has long been agreed upon

O

Open societies—societies with inbuilt class mobility (or at least the myth of such)

Organic analogy—society is like an organism, with differentiated systems held together by mutual interdependence; it will evolve over time from being simple and homogeneous (like an amoeba) to being heterogeneous and complex (also sometimes called organismic analogy)

Out of Africa model—holds that *Homo sapiens* spread globally in waves of migration outward from East Africa; see also 'single origin model' (contrast with 'multiple origins model')

P

Pandemic—disease that spreads around the world with great speed

Paradigm shift—a radical transition in thinking; see also 'paradigm'

Paradigm—a theoretical framework that guides scientists in deciding what questions to ask, and forms a lens through which scientists interpret data

Participant-observation—a hands on approach to ethnographic data collection that involves living among members of the culture under study, speaking their language, and participating in as much of their daily lives as is possible.

Passive immunity—immunity acquired by infants from breastfeeding; antibodies are passed along via breast milk

Pastoralism—herding lifestyle; nomadic or semi-nomadic.

Pathogen—germ or infectious agent

Pathogenesis—the creation or genesis of damage or pathology in the body

Patrilineage—descent group in which heredity is figured through the male lineage

Peppered moths—species of moth found most famously in the forests of England; its coloration can evolve from lighter to darker and back again when the environment changes so that one coloration is more adaptive than the other

Performative—a communicative act that does not just describe a situation but creates it, as when pronouncing a couple married makes them married

Phagocytes—those leukocytes or white blood cells that eat or otherwise get rid of pathogens

Phenotype—the manifestation or measurable expression of the genotype

Phytochemicals—components or chemical compounds in a plant; can provide ingrown or natural protection from predators and disease

Placebo effect—a measurable, observable, or felt improvement in health that is not directly attributable to biomedical treatment; see also 'meaning response'

Placebo—an inert substance or an act that is not biomedical and does not effect a clinical cure

Plasticity—malleability, particularly in relation to an organism's developmental range

Political economy—a theoretical perspective that considers phenomena or processes in political and economic context; rather than to focus on proximate factors, the focus is on ultimate causes

Polygenic—traits that develop through the interaction of a collection of numerous genes (poly means many)

Polymorphic—having various forms (poly means many); see also 'alleles'

Porotic hyperostosis—bones that are more porous than they should be; a sign of iron deficiency, which (whether caused by malnutrition or infection) triggers the expansion of bone tissues that form red blood cells

Primary health care—a public health movement that focuses on preventing problems before they happen.

Proximate—immediate; very near; proximate causes are direct causes

Q

Quickening—the point in gestation at which fetal movement can be felt by the birth mother

R

Race—a clearly differentiated sub-species or sub-group within a species; races have features that regularly co-occur; human sub-species do not exist

Racialization—the simple classification of people according to so-called race; non-evaluative (contrast with 'racism')

Racism—an evaluative position in which people are classed by so-called race and the races are rated and ranked (contrast to 'racialization')

Recessive allele—gene form that can only be expressed if paired with a self-same allele; can be dominated or overridden by a dominant allele; a recessive trait is a trait overridden in this way

Reciprocal relationship—an exchange relationship infused with emotional value, and meant to be of long-standing duration

Red Queen Hypothesis—the idea that a population in which genetic variation is not maintained is at a distinct disadvantage in a world that is constantly producing new environmental pressures

Redistribution—goods are taken from one subgroup for use with another, or for use society-wide (e.g., to build roads)

Reductionism—the idea that an entire system can be explained by a single aspect of that system

Relativism—evaluates the ideas and practices enacted by members of a given culture by that culture's own standards

Replacement hypothesis—holds that populations of *Homo sapiens* who left Africa in earlier waves (see 'out of Africa model') were decimated by diseases newer population waves brought, or were driven out or otherwise replaced by the newcomers

Reservoir host—living organism that carries a pathogen as a host and serves as a vector or pathogen-carrying organism

Resilience—strength in the face of otherwise stressful situations

Resistance—ability of an organism to withstand a particular pathogen or stressor; in pathogens, an evolved capacity to remain unaffected by an antibiotic due to vulnerable organisms dying off prior to reproducing themselves

Rite of passage—a set of ritual acts intended to move a person or a group of people from one social status to the next

Ritual—set of actions performed for their culturally relevant symbolic value; a multimedia event that, when enacted, increases group cohesion

RNA (ribonucleic acid)—biological molecule known for its role in protein synthesis; also found to regulate gene expression

Role incongruity—when one is not living up to the expectations entailed in the role one has been placed in or has elected to take on

S

Scavenging—minimum-impact lifestyle in which subsistence depends on picking up dead organisms and fruit (etc.) that has dropped off the plant

Seasonality—when changing seasons lead to shifts in diet quality and quantity

Sedentism—when a population lives in a permanent settlement, as agriculturalists do

Sex—biologically differentiated status of male or female, related to genital and chromosomal endowment

Sexual reproduction—reproduction or production of offspring in which more than one organism contributes genes, and so genes are shuffled around, generally with half coming from one parent and half from one other; ensures variation, as every offspring can receive a slightly different combination of genes

Sexual selection—selective mating, driven by unconscious impulses or conscious rules; happens in human populations for instance when kinship or class-related rules or ethnic boundaries limit who can build a family with whom; see also 'non-random mating'

Sickle cell anemia—gene-linked disease causing red blood cells to carry less oxygen and to form an abnormal crescent shape that can clog circulation; generally incurable

Simple society—band or tribe organization; homogenous in structure (refers to social organization, not culture)

Smallpox—potentially fatal, frequently epidemic acute viral infection accompanied by an extensive pustular or blister-like rash and high fever; spread via contact with pustular fluid (pustules can be near or in mouth or nose, so that droplets in an infected person's breath also may lead to contagion)

Social cohesion—when a society holds itself together

Social condensation—when larger groups break into smaller factions, much as water vapor will condense into droplets on a cold drink bottle

Social justice—the equitable distribution of basic human rights such as the right to healthful living conditions, and equal opportunities for equal outcomes among all social groups

Social soundness—when aid groups work with community leaders and members to create a good fit between programs and cultures

Social stratification—hierarchical social organization

Social structure—refers to the way in which a society is structured or built from the relationships between the classes or groups of people within it

Somatization—the projection of mental attitudes or concerns onto the body so that they are expressed as physical symptoms; often this provides an outlet or channel for stress

Specialization—when different groups within a society do different jobs and are therefore interdependent

Species—a discrete organism type that can reproduce itself, for instance by interbreeding and giving birth to fertile offspring; reproduction of fertile offspring is impossible across species boundaries

State—hierarchical socio-political entity in which bureaucracy is centralized, with multiple parts, including a formal legal system

Status inconsistency—when others view a person in a way that is inconsistent with the way she or he expects to be viewed; when a person is not ascribed the status she or he feels that she or he deserves

Stigmatized—marked for rejection, generally with a visual sign (e.g., a tattoo on a visible part of one's body, a sign on one's clothing)

Stress—the body's immediate response to environmental pressures; includes reactions of the nervous, hormonal, and immune systems; protective in the short run; key biochemicals entailed are cortisol and catecholamines, including epinephrine; see also 'fight or flight response' (regarding chronic stress, see 'allostasis'; contrast with 'environmental pressure')

Structural violence—occurs when the shape of a given social structure harms or is harmful to the people who occupy certain positions within that social structure

Subsistence strategy—approach to or means of making a living or extracting food from the environment, such as by foraging or industrial agriculture

Surplus—extra food or material items that can be stored

Surrogate—stand in mother or father for procreative purposes; surrogate mothers (gestational surrogates) can carry (gestate) children for their legal mothers.

Syndemic—a cluster of health problems that work together synergystically, reinforcing and often exacerbating each other

Systems thinking—a point of view that highlights relationships and what emerges from them; properties of a system can be neither explained nor determined by examining its parts alone

T

Taboo—rule against coming in contact with something, often due to it being a well of power; see also 'anomaly'

Transgender—a person who crosses or straddles the border between male and female (while androgyny focuses on a blend, transgender roles in some ways leave our two-gender system behind)

Transmembrane conductance regulators (TRs)—biological structures that move (conduct, transfer) things across membranes, such as the intestinal membrane, regulating their movement or flow across these membranes; certain TRs are implicated in resistance and susceptibility to cholera and in cystic fibrosis

Tribe—egalitarian socio-political group made up of a number of bands that have have cross-cutting inter-band ties; leadership is achievement based

Tuberculosis—potentially fatal bacterial infection of the lungs; can spread to other organs and the skeleton; long-lasting; spread via droplets released when a sick person coughs or sneezes

U

Ultimate—in the end; ultimate causes are where the buck stops

Unilineal evolution—the idea that culture (or anything) evolves in one direction over time, progressively improving (non-improvement would be 'devolution')

Upper Paleolithic Revolution—dramatic shift in the archeological record occurring about 75,000 years ago indicating onset of true behavioral modernity in *Homo sapiens sapiens,* including full-fledged capacity for culturally supported cooperative social life; see also 'Cultural Revolution'

V

Vector—organism that carries a pathogen to another organism (e.g., mosquito)

Z

Zoonotic—a disease that originated in an animal species and then jumped to humans

References

American Anthropological Association. 2007. "Only Skin Deep." In *Race: Are We So Different?* www.understandingrace.org/humvar/skin_01.html.

American Anthropological Association. 2007. "Global Census." In *Race: Are We So Different?* www.understandingrace.org/lived/global_census.html.

Arbib, Michael A., Katja Liebal, and Simone Pika. "Primate Vocalization, Gesture, and the Evolution of Human Language." *Current Anthropology* 49, no. 6 (2008): 1053–76.

Archibold, Randal C. "Indians' Water Rights Give Hope for Better Health." *New York Times*, August 30, 2008, A20.

Balée, William. *Inside Cultures: A New Introduction to Cultural Anthropology.* Walnut Creek, CA: Left Coast Press, Inc., 2012.

Barr, D. *Health Disparities in the United States: Social Class, Race, Ethnicity, and Health.* Baltimore, MD: Johns Hopkins University Press, 2008.

Bart, P. "Why Women's Status Changes with Middle Age." In *Sociological Symposium*. Blacksburg, VA: Virginia Polytechnic Institute and State University, Department of Sociology, 1969.

Bateson, Patrick, David Barker, Timothy Clutton-Brock, Debal Deb, Bruno D'Udine, Robert A. Foley, Peter Gluckman, et al. "Developmental Plasticity and Human Health." *Nature* 430 (2004): 419–21.

Becker, Anne E. *Body, Self, and Society: The View from Fiji.* Philadelphia: University of Pennsylvania Press, 1995.

Becker, G., A. Butler, and R. Nachtigall. "Resemblance Talk: A Challenge for Parents Whose Children Were Conceived with Donor Gametes in the US." *Social Science and Medicine* 61 (2005): 1300–09.

Belkin, Douglas. "Blue Eyes Are Increasingly Rare in the Americas." *New York Times*, October 18, 2006.

Bilu, Yoram. "From Milah (Circumcision) to Milah (Word): Male Identity and Rituals of Childhood in the Jewish Ultraorthodox Community." *Ethos* 31, no. 2 (2003): 172–203.

Biodynamic Farming and Gardening Association. "About Biodynamics." www.bio-dynamics.com/biodynamics.

Blackwell, B, S.S. Bloomfield, and C.R. Buncher. "Demonstration to Medical Students of Placebo Responses and Non-Drug Factors." *Lancet* 1, no. 7763 (1972): 1279–82.

Blakeslee, Sandra. "Cells That Read Minds." *New York Times*, January 10, 2006.

Bogin, Barry. "The Tall and the Short of It." *Discover* 19, no. 2 (1998): 40–44.

Bruner, Emiliano, and Giorgio Manzi. "Variability in Facial Size and Shape among North and East African Human Populations." *Italian Journal of Zoology* 71, no. 1 (2004): 51–56.

Burton, John W. *Culture and the Human Body*. Long Grove, IL: Waveland Press, 2001.

California Newsreel. *Finding Hope for the Future by Reclaiming the Past*. San Francisco: California Newsreel, 2008a.

———. *Unnatural Causes: In Sickness and in Wealth*. San Francisco: California Newsreel with Vital Pictures, 2008b.

———. *Unnatural Causes: When the Bough Breaks*. San Francisco: California Newsreel with Vital Pictures, 2008c.

Cannon, Walter B. ""Voodoo" Death." *American Anthropologist* 44, no. 2 (1942): 169–81.

Cassidy, C. "The Good Body: When Bigger Is Better." *Medical Anthropology* 13 (1991): 181–213.

Cohen, Mark Nathan. *Health and the Rise of Civilization*. New Haven, CT: Yale University Press, 1989.

Cormier, Loretta A. *The Ten-Thousand Year Fever: Rethinking Human and Wild-Primate Malarias*. Walnut Creek, CA: Left Coast Press, Inc., 2011.

Cousins, Norman. *Anatomy of an Illness as Perceived by the Patient: Reflections on Healing and Regeneration*. New York: Bantam Books, 1981.

Dailey, Kate, and Abby Ellin. "America's War on the Overweight." *Newsweek* 2009.

Darwin, Charles R. *On the Origin of Species by Means of Natural Selection, or the Preservation of Favoured Races in the Struggle for Life*. 1st ed. London: John Murray, 1859.

David, Richard J., and James W. Collins. "Differing Birth Weight among Infants of U.S.-Born Blacks, African-Born Blacks, and U.S.-Born Whites." *New England Journal of Medicine* 337, no. 17 (1997): 1209–14.

Diamond, Jared. *Guns, Germs, and Steel: The Fates of Human Societies*. New York: WW Norton and Company, 1997.

———. "Race without Color." *Discover* 15, no. 11 (1994): 82–89.

Doleac, Jennifer L., and Luke C. D. Stein. 2010. "The Visible Hand: Race and Online Market Outcomes." ssrn.com/abstract=1615149.

Douglas, Mary. *Purity and Danger: An Analysis of Concepts of Pollution and Taboo*. London: Routledge and Keegan Paul, 1966.

Dressler, William W., and James R. Bindon. "The Health Consequences of Cultural Consonance: Cultural Dimensions of Lifestyle, Social Support and Blood Pressure in an African American Community." *American Anthropologist* 102, no. 2 (2000): 244–60.

Dretzin, Rachel. "Interview—Clifford Nass." Boston: WGBH Educational Foundation. 2010. www.Pbs.Org/Wgbh/Pages/Frontline/Digitalnation/Interviews/Nass.Html.

Drexler, Madeline. "How Racism Hurts—Literally." Boston Globe, July 15, 2007.

Dunbar, Robin. *Grooming, Gossip, and the Evolution of Language*. Cambridge, MA: Harvard University Press, 1996.

Durkheim, Emile. *The Division of Labor in Society*. Translated by Lewis A. Coser. New York: Free Press, 1997 (1893).

Eaton, S.B., S.B. Eaton III, and M.J. Konner. "Paleolithic Nutrition Revisited: A Twelve-Year Retrospective on Its Nature and Implications." *European Journal of Clinical Nutrition* 51 (1997): 207–16.

Economist. "Eyeing up the Collaboration." *Economist* 381, no. 8502 (2006): 90.

———. "I Am Just a Poor Boy Though My Story's Seldom Told: Neuroscience and Social Deprivation." *Economist* 391, no. 8625 (2009): 82.

———. "Mens Sana in Corpore Sano." *Economist* 396, no. 8689 (2010a): 75–76.

———. "Really New Advances—RNA." *Economist* 383, no. 8533 (2007): 87–89.

———. "Stiletto Stiffness." *Economist* 396, no. 8691 (2010b): 84.

Ehrlich, Paul R. *Human Natures*. New York: Penguin Books, 2006.

Engels. "Manchester in 1844 (Excerpted from the Condition of the Working Class in England in 1844)." In *Essential Works of Socialism*, edited by Irving Howe, 58–71. Clinton, MA: Colonial Press, 1970 [1845].

Eppig, Christopher, Corey L. Fincher, and Randy Thornhill. "Parasite Prevalence and the Worldwide Distribution of Cognitive Ability." *Proceedings of the Royal Society B* (2010).

Farmer, Paul. *Pathologies of Power: Health, Human Rights, and the New War on the Poor*. Los Angeles: University of California Press, 2005.

Flint, M. "The Menopause: Reward or Punishment?" *Psychosomatics* 16 (1975): 161–63.

Fluehr-Lobban, Carolyn. *Race and Racism: An Introduction*. Lanham, MD: AltaMira, 2006.

Fuentes, Agustin. *Evolution of Human Behavior*. New York: Oxford University Press, 2009.

Galanti, Geri-Ann. *Caring for Patients from Different Cultures: Case Studies from American Hospitals*. 4th ed. Philadelphia: University of Pennsylvania Press, 2008.

Galtung, Johan. "Violence, Peace, and Peace Research." *Journal of Peace Research* 6, no. 3 (1969): 167–91.

Garcia, Richard. "The Misuse of Race in Medical Diagnosis." *Chronicle of Higher Education* 49, no. 35 (2003): B15.

Gersten, Omer. "The Path Traveled and the Path Ahead for the Allostatic Framework: A Rejoinder on the Framework's Importance and the Need for Further Work Related to Theory, Data, and Measurement. " *Social Science & Medicine* 66, no. 3 (2008): 531–35.

Gill, G. W. "Does Race Exist? A Proponent's Perspective." *Nova* 2000. www.pbs.org/wgbh/nova/first/gill.html.

Gilman, Sander L. *Making the Body Beautiful: A Cultural History of Aesthetic Surgery.* Princeton, NJ: Princeton University Press, 1999.

Goldschmidt, Walter. *The Bridge to Humanity: How Affect Hunger Trumps the Selfish Gene.* New York: Oxford University Press, 2006.

Goodman, Alan H., and Thomas L. Leatherman, eds. *Building a New Biocultural Synthesis: Political Economic Perspectives on Human Biology.* Ann Arbor: University of Michigan Press, 1998.

Goodyear, Dana. "Grub: Must We Eat Insects?" *New Yorker* August 15 and 22, no. 20B (2011): 38–47.

Gould, Stephen Jay. "The Geometer of Race." *Discover* 15, no. 11 (1994): 65–69.

Gross, Daniel R. *Discovering Anthropology.* Mountain View, CA: Mayfield, 1992.

Harron, Mary. "A Hero for Our Times." *Independent,* January 29, 1995.

Herdt, Gilbert. *Guardians of the Flutes: Volume 1. Idioms of Masculinity.* Chicago: University of Chicago Press, 1994 [1981].

Jhally, Sut. *Tough Guise: Violence, Media, and the Crisis in Masculinity.* Northampton, MA: Media Education Foundation, 1999.

Joralemon, Donald. *Exploring Medical Anthropology.* 2nd ed. San Francisco: Pearson Allyn and Bacon, 2006.

Katz, J. N. *The Invention of Hetereosexuality.* Chicago: University of Chicago Press, 2007 [1995].

Keckley, Paul H., and Howard R. Underwood. *Medical Tourism: Consumers in Search of Value.* Washington, DC: Deloitte Center for Health Solutions, 2008.

Kidder, Tracy. *Mountains Beyond Mountains: The Quest of Dr. Paul Farmer, a Man Who Would Cure the World.* New York: Random House, 2003.

Kluger, Jeffrey. "The Science of Appetite." Time, May 31, 2007.

Kristof, Nicholas D. "Raising the World's IQ." *New York Times,* November 4, 2008, A43.

Kuhn, Thomas S. *The Structure of Scientific Revolutions.* Chicago: University of Chicago Press, 1996 [1962].

Kumar, Pooja. "Indian Surrogacy: Blessing, Business or Both?" *Neem Magazine* 1, no. 4 (2008). www.neemmagazine.com/culture63.

LaFleur, William R. *Liquid Life: Abortion and Buddhism in Japan* Princeton, NJ: Princeton University Press, 1994.

Langness, L L. *The Study of Culture.* 3rd, revised ed. Novato, CA: Chandler and Sharp, 2005.

Larsen, Clark Spencer. "The Agricultural Revolution as Environmental Catastrophe: Implications for Health and Lifestyle in the Holocene." *Quaternary International* 150, no. 1 (2006): 12–20.

Leach, Edmund R. "Time and False Noses." In *Reader in Comparative Religion: An Anthropological Approach,* edited by William A. Lessa and Evon Z. Vogt, 226-229. San Francisco: Harper and Row, 1979 [1964].

Lee, Richard B. "What Hunters Do for a Living, or, How to Make out on Scarce Resources." In *Nutritional Anthropology,* edited by Alan H. Goodman, Darna

L. Dufour and Gretel H. Pelto, 35–46. Mountain View, CA: Mayfield, 2000 [1968].

Lenkeit, Roberta Edwards. *Introducing Cultural Anthropology.* 4th ed. San Francisco: McGraw Hill, 2009.

Lewontin, Richard. "The Apportionment of Human Diversity." *Evolutionary Biology* 6 (1972): 391–98.

MacDorman, M. F., and T. J Mathews. *Recent Trends in Infant Mortality in the United States. NCHS Data Brief, No 9.* Hyattsville, MD: National Center for Health Statistics, 2008.

Maxted, Ian. 2000. "Cider and Eighteenth-Century Evidence-Based Healthcare." *Devon's Heritage,* Devon Library Services. www.devon.gov.uk/localstudies/100676/1.html.

Mays, V.M., S.D. Cochran, and N.W. Barnes. "Race, Race-Based Discrimination, and Health Outcomes among African Americans." *Annual Review of Psychology* 58 (2007): 201–25.

McDowell, M. A., C. D. Fryar, C. L. Ogden, and K. M. Flegal. "Anthropometric Reference Data for Children and Adults: United States, 2003–2006." In *National Health Statistics Reports.* Hyattsville, MD: National Center for Health Statistics, 2008.

McElroy, Ann, and Patricia K. Townsend. *Medical Anthropology in Ecological Perspective.* 4th ed. Boulder, CO: Westview Press, 2004.

McKenna, James J., and Thomas McDade. "Why Babies Should Never Sleep Alone: A Review of the Co-Sleeping Controversy in Relation to Sids, Bedsharing and Breast Feeding." *Paediatric Respiratory Reviews* 6 (2005): 134–52.

Mead, Margaret. *Male and Female.* New York: HarperCollins, 2001 [1949].

———. *Sex and Temperament in Three Primitive Societies.* New York: HarperCollins, 2001 [1935].

Meigs, Anna S. "Blood Kin and Food Kin." In *Conformity and Conflict*, edited by James P. Spradley and David W. McCurdy, 117–24. Boston: Little, Brown and Company, 1987.

———. "Food as a Cultural Construction." In *Food and Culture: A Reader*, edited by C. Counihan and P. Van Esterik, 95–106. New Brunswick, NJ: Rutgers University Press, 1997.

Mennella, Julie A., Coren P. Jagnow, and Gary K. Beauchamp. "Prenatal and Postnatal Flavor Learning by Human Infants." *Pediatrics* 107, no. 6 (2001): e88.

Moerman, Daniel. *Meaning, Medicine, and the 'Placebo Effect.'* Cambridge: University of Cambridge Press, 2002.

Mullin, S., and M. Duwell. "Despite much progress, significant disparities persist in New York City: New report details disproportionate burden of illness and premature death among poor, as well as Black and Hispanic New Yorkers." Press Release (2004). New York City Department of Health and Mental Hygiene Office of Communications. www.nyc.gov/html/doh/html/press_archive04/pr085-0715.shtml.

National Institute of Diabetes and Digestive and Kidney Diseases. *The Pima Indians: Pathfinders for Health.* Bethesda, MD: National Institute of Diabetes and Digestive and Kidney Diseases, National Institutes of Health, 2002.

Navdanya. 2009. "Navdanya: Two Decades of Service to the Earth and Small Farmers." www.Navdanya.Org/Attachments/Navdanya.Pdf.

Nisbett, Richard E., and Takahiko Masuda. "Culture and Point of View." *Proceedings of the National Academy of Sciences* 100, no. 19 (2003): 11163–70.

Nova ScienceNOW. 2007. "A Tale of Two Mice." NOVA/WGBH Educational Foundation. www.pbs.org/wgbh/nova/sciencenow/3411/02-tale-flash.html.

Odling-Smee, F. John, Kevin N. Laland, and Marcus W. Feldman. *Niche Construction: The Neglected Process in Evolution.* Princeton, NJ: Princeton University Press, 2003.

Park, Alice. "Can Sugar Substitutes Make You Fat?" *Time,* 2008. www.Time.Com/Time/Health/Article/0,8599,1711763,00.Html.

Pembrey, Marcus E, Lars Olov Bygren, Gunnar Kaati, Sören Edvinsson, Kate Northstone, Michael Sjöström, Jean Golding, and The ALSPAC Study Team. "Sex-Specific, Male-Line Transgenerational Responses in Humans." *European Journal of Human Genetics* 14, no. 2 (2006): 159–66.

Perry, Richard J. *Five Key Concepts in Anthropological Thinking.* Upper Saddle River, NJ: Pearson Education, 2003.

Plsek, Paul. "Appendix B: Redesigning Health Care with Insights from the Science of Complex Adaptive Systems." In *Crossing the Quality Chasm: A New Health System for the 21st Century,* edited by Institute of Medicine Committee on Quality of Health Care in America, 309–22. Washington, DC: National Academy Press, 2001.

Pollan, Michael. *The Omnivore's Dilemma: A Natural History of Four Meals.* New York: Penguin Books, 2006.

Popenoe , Rebecca. *Feeding Desire: Fatness, Beauty, and Sexuality among a Saharan People.* London: Routledge, 2004.

Public Broadcasting System, PBS. *The Evolutionary Arms Race.* Boston: WGBH Educational Foundation and Clear Blue Sky Productions, 2001.

Raymond, J. "GPS Addict? It May Be Eroding Your Brain." *Today Health,* NBC News 2010. today.msnbc.msn.com/id/40138522/ns/health-mental_health/#.UClgvaOs_5E.

Relethford, J. H. "Human Skin Color Diversity Is Highest in Sub-Saharan African Populations." *Human Biology* 72, no. 5 (2000): 773–80.

Rich, Stephen M., Fabian H. Leendertz, Guang Xua, Matthew LeBreton, Cyrille F. Djoko, Akoah N. Aminake, Eric E. Takang, et al. "The Origin of Malignant Malaria." *Proceedings of the National Academy of Sciences* 106, no. 35 (2009): 14902–07.

Ridley, Matt. *The Agile Gene.* New York: Perennial, 2003.

———. *Genome: The Autobiography of a Species in 23 Chapters.* New York: Harper Perennial, 1999.

Riis, Jacob. *How the Other Half Lives: Studies among the Tenements of New York*. Whitefish, MT: Kessinger, 2004 [1890].

Roscoe, Will. *The Zuni Man-Woman*. Albuquerque: University of New Mexico Press, 1992.

Rubel, Arthur J., Carl W. O'Nell, and Rolando Collado-Ardón. *Susto: A Folk Illness*. Los Angeles: University of California Press, 1984.

Sanders, Clinton R. "Marks of Mischief. Becoming and Being Tattooed." *Journal of Contemporary Ethnography* 16, no. 4 (1988): 395–432.

Schmidt, D. "The Family as the Unit of Medical Care." *Journal of Family Practice* 7, no. 2 (1978): 303–13.

Schor, Juliet B. *The Overworked American: The Unexpected Decline of Leisure*. New York: Basic Books, 1993.

Shenk, David. *The Genius in All of Us: New Insights into Genetics, Talent, and IQ*. New York: Anchor Books, 2011.

Simmons, Ann M. "Where Fat Is a Mark of Beauty." *Los Angeles Times*, September 30, 1998.

Sobo, Elisa J. "Medical Travel: What It Means and Why It Matters." *Medical Anthropology* 29, no. 1 (2009): 326–35.

———. *One Blood: The Jamaican Body*. Albany: State University of New York Press, 1993.

———, and Martha Loustaunau. *The Cultural Context of Health, Illness and Medicine*. 2nd ed. Santa Barbara, CA: Praeger, 2010.

Spencer, Herbert *Principles of Sociology*, 3 vols. Piscataway, NJ: Transaction, 2002 [1874–1896].

Steiner, Rudolf. *The Philosophy of Freedom: The Basis for a Modern World Conception*. Trans. Michael Wilson. Forest Row, UK: Rudolf Steiner Press, 2008 [1894].

———. *Study of Man (General Education Course): Fourteen Lectures Given in Stuttgart between 21 August and 5 September 1919*. Translated by Daphne Harwood and Helen Fox. Forest Row, UK: Rudolf Steiner Press, 2007.

Strom, Stephanie. "Has 'Organic' Been Oversized?" *New York Times*, July 8, 2012, BU1.

Swarns, Rachel L. "DNA Tests Offer Immigrants Hope or Despair." *New York Times*, April 10, 2007, A1.

Thomas, Anthony. "Fat." In *Frontline*, Season 16, Episode 17, 1998.

Tohono O'odham Community Action. 2010. www.tocaonline.org.

Tomasello, Michael. "Human Culture in Evolutionary Perspective." In *Advances in Culture and Psychology*, edited by M. Gelfand, 5–51. New York: Oxford University Press, 2011.

Trevathan, Wenda R. "Evolutionary Obstetrics." In *Evolutionary Medicine*, edited by Wenda R. Trevathan, E.O. Smith, and James J. McKenna, 183–207. New York: Oxford University Press, 1999.

Turner, Victor W. "Betwixt and Between: The Liminal Period in 'Rites De Passage.'" In *Reader in Comparative Religion: An Anthropological Approach*, edited by William A. Lessa and Evon Z. Vogt, 234–43. San Francisco: Harper and Row, 1979 [1964].

Ulrich, Roger S. "Evidence-Based Healthcare Architecture." *Lancet* 368, Supplement 1 (2006): S38–S39.

———. "View through a Window May Influence Recovery from Surgery." *Science* 224, no. 4647 (1984): 420–21.

Van Gennep, Arnold. *The Rites of Passage*. Translated by Monika B. Vizedom and Gabrielle L. Caffee. Chicago: University of Chicago Press, 1960 [1908].

Waber, R. L., B. Shiv, Z. Carmon, and D. Ariely. "Commercial Features of Placebo and Therapeutic Efficacy." *Journal of the American Medical Association* 299, no. 9 (2008): 1016–17.

Waddington, C. H. "The Epigenotype." *Endeavour* 1 (1942): 18–20.

Wade, Nicholas. "Human Culture, an Evolutionary Force." *New York Times,* March 2, 2010, D1.

Watkins, A. "Contemporary Context of Complementary and Alternative Medicine: Integrated Mind-Body Medicine." In *Fundamentals of Complementary and Alternative Medicine*, edited by M. Micozzi, 49–63. New York: Churchill Livingstone, 1996.

Weller, Susan C., Trenton K. Ruebush II, and Robert E. Klein. "An Epidemiological Description of a Folk Illness: A Study of Empacho in Guatemala." In *Anthropological Approaches to the Study of Ethnomedicine*, edited by Mark Nichter, 19-31. New York: Gordon and Breach, 1992.

Whitelaw, Emma. "Epigenetics: Sins of the Fathers, and Their Fathers." *European Journal of Human Genetics* 14, no. 2 (2006): 131–32.

Whiting, John W. M., and Irvin L. Child. *Child Training and Personality: A Cross-Cultural Study*. New Haven, CT: Yale University Press, 1953.

Wilder, Billy. "Some Like It Hot." United Artists, 1959.

Wyatt, S. B., D. R. Williams, R. Calvin, F. C. Henderson, E. R. Walker, and K. Winters. "Racism and Cardiovascular Disease in African Americans." *American Journal of the Medical Sciences* 325, no. 6 (2003): 315–31.

Yuki, Masaki, William W. Maddux, and Takahiko Masuda. "Are the Windows to the Soul the Same in the East and West? Cultural Differences in Using the Eyes and Mouth as Cues to Recognize Emotions in Japan and the United States." *Journal of Experimental Social Psychology* 43, no. 2 (2007): 303–11.

Zernike, Kate. "Sizing up America: Signs of Expansion." *New York Times Magazine* March 1, 2004.

Index

About the Author

Elisa (EJ) Sobo, Professor of Anthropology at San Diego State University, teaches *The Dynamics of Biocultural Diversity* as well as courses on topics such as medical anthropology and childhood. Dr. Sobo is presently on the editorial boards of *Anthropology & Medicine* and *Medical Anthropology*, and she is book reviews editor for *Medical Anthropology Quarterly*. She has served as an elected member of the executive board of the Society for Medical Anthropology as well as on the medical committee of the Royal Anthropological Institute in the UK, and she now serves as co-chair of the Committee on Public Policy of the American Anthropological Association. Dr. Sobo has written numerous peer-reviewed journal articles and has authored, co-authored, and co-edited eleven previous books on various topics relevant to our biocultural experience. Her current research examines cultural models of child development as applied in classroom teaching, particularly in the Waldorf or Steiner education system.